DIET
AND
DRUG
INTERACTIONS

DIET
and
DRUG
INTERACTIONS

Daphne A. Roe, M.D., F.R.C.P.

Professor of Nutrition
Division of Nutritional Sciences
Cornell University
Ithaca, New York

An **avi** Book
Published by Van Nostrand Reinhold
New York

An AVI Book
(AVI is an imprint of Van Nostrand Reinhold)

Printed in the United States of America

Designed by Monika Grejniec

Van Nostrand Reinhold
115 Fifth Avenue
New York, New York 10003

Van Nostrand Reinhold International Company Limited
11 New Fetter Lane
London EC4P 4EE, England

Van Nostrand Reinhold
480 La Trobe Street
Melbourne, Victoria 3000, Australia

Macmillan of Canada
Division of Canada Publishing Corporation
164 Commander Boulevard
Agincourt, Ontario M1S 3C7, Canada

16 15 14 13 12 11 10 9 8 7 6 5 4 3 2

Library of Congress Cataloging-in-Publication Data

Roe, Daphne A., 1923–
    Diet and drug interactions / Daphne A. Roe.
      p. cm.
    "An AVI Book."
    Includes bibliographies and index.
    ISBN 0-442-20487-6 :
    1. Drug–nutrient interactions.   I. Title.
    [DNLM: 1. Drug Interactions.   2. Food.   3. Nutrition—drug
effects.   QV 38 R698d]
RM302.4.R624 1988
615'.7045—dc19
DNLM/DLC
for Library of Congress   88-5669

Dedicated to my husband, "Shad" Roe, who offered me so much useful advice as I sat beside my word processor preparing this manuscript.

I would like to express my thanks to my secretary, Beverly Hastings, for her invaluable assistance.

CONTENTS

PREFACE

When we learn from a patient, clinician, or medical record that a drug has been discontinued, it is logical to ask why. The drug may no longer be needed; it may not have produced the desired effect; it may have produced an adverse reaction; a better drug may be available to replace the original drug. The patient may have discontinued the drug because he or she could not see why it was necessary; or the patient may have discontinued the drug because of unpleasant side effects.

A drug may not work because its absorption is reduced by physical or chemical interaction with another drug or a food component. It may also not work because the patient's metabolism is speeded up or inhibited to an extent such that the desired duration of drug action is not obtained. Such an effect may be related to a change in diet.

Side effects may be related to consumption of specific foods or beverages or to an overall change in nutritional status. Drug–food and drug–alcohol incompatibility reactions are frequent but are avoidable if a patient is warned of their possible occurrence. Drugs may also produce nutritional deficiencies, especially in a patient whose diet is marginal in those nutrients depleted by the particular drug. Careful prescribing practices together with appropriate nutrient supplements will serve to reduce the risk of these incompatibilities.

This book is addressed not only to physicians, but also to other health care providers who are responsible for instructing patients on safe and effective usage of medications. Since readers may be either MDs, physician's assistants, nurses, pharmacists, or dietitians, a common language has been adopted with deliberate omission of the jargon of any one profession. Information is given as briefly as possible on the mechanisms responsible for food and formula interference with drug absorption and metabolism, on compatibility reactions, on effects of drugs on nutritional status, and of nutritional status on drug disposition. Food intolerances resulting from intentional additives in drugs and foods are described; up-to-date tables are provided, listing drug and dietary sources of these food chemicals. Diets free of such substances and therefore tolerated by patients who have such reactions are also given.

The use and abuse of vitamins as supplements and as drugs is reviewed with particular emphasis on nutrient toxicity and the effects of chronic vitamin overload.

The risk of drug–nutrient interactions and their adverse outcomes can be predicted using expert systems and knowledge bases that include attributes of drugs, drug users, and drug regimens.

Guidelines are provided on when and how best to take the commonly prescribed drugs in order to avoid dietary interference and to minimize adverse effects related to foods, beverages, or change in nutritional status.

Drug interference with the assessment of nutritional status has been subdivided into analytical and biological interferences and is presented with an up-to-date bibliography.

Since this manual is intended for use in medical offices, clinics, and hospitals, whether in the ward, dietary department, or pharmacy, its contents include information that can supply members of the health care team with updates on the drug and diet interactions they are likely to meet and indicate when and how adverse outcomes of these interactions are preventable.

This book is for ready reference. It is most sincerely hoped that its use will serve to improve drug compliance and to reduce adverse drug reactions.

Chapter 1

DEFINITION, DIAGNOSIS, AND RISK OF
DIET AND DRUG INTERACTIONS

In describing diet and drug interactions, the term *diet* is used to include foods, nutrients, or nonnutrient components of food; nutrient formulas and their constituents; and nutrient supplements. The term *drug* includes alcohol and other social drugs as well as therapeutic drugs. Diet and drug interactions include all interactions occurring in vitro (in food or drug products), or in vivo (in the body) that affect the stability or disposition of food or drug components, whether or not they measurably affect physiological function or health.

TYPES OF DIET AND DRUG INTERACTIONS

Drug and diet interactions are of four main types:

1. Physicochemical or Chemical Interactions in vitro. These include chemical and photolytic interactions between chemical additives and nutrients in food, formula, or drug products. The destruction of thiamin by sulfites (Gubler 1984) is an example of an interaction between an intentional food additive and a nutrient. Photochemical interaction between nutrients (where one behaves as a phototoxic substance and potentiates the photodegradation of another nutrient) is observed when riboflavin-containing formula solutions intended for intravenous use are exposed to light. Riboflavin can be phototoxic in the mixture and can potentiate the photodegradation of other nutrients present including amino acids and carotenoids (Roe in press).

2. Physicochemical Interactions in vivo. These occur in the gastrointestinal tract. Drug particles can adsorb nutrients, drugs can adsorb onto dietary fiber sources, chelates can form between divalent cations and drugs such as tetracycline, and drugs and nutrients can form precipitates (Gibaldi 1977).

1

3. Metabolic Interactions. Nutrients or nonnutrients in food alter the rate of drug metabolism; drugs likewise can alter the rate (or extent) of metabolism of nutrients. The former type of interaction operates when a high-protein diet speeds up the rate of metabolism of the drug theophylline (Kappas et al. 1976). Metabolic interactions that involve effects of drugs on nutrient disposition include those that alter rates of synthesis of the active forms of vitamins. For example, alcohol ingestion impairs the synthesis of flavin coenzymes (Rosenthal et al. 1973).

Drugs sometimes inhibit enzymes required for the metabolism of nonnutrient food components or alcohol. Examples include effects of monamine oxidase inhibitor drugs on the metabolism of tyramine and effects of disulfiram on the metabolism of acetaldehyde (Asatoor et al. 1963; Morgan and Cagan 1974).

4. Functional Interactions. Both the rate and the extent of elimination of drugs and nutrients may be altered by tissue injury, such as to the kidneys, where impaired renal function may be due to the feeding of high-protein diets (Hostetter 1984). Nutrient depletion can result from cellular injury due to the effects of drugs on the intestinal mucosa. Examples include the drug-induced enteropathies caused by neomycin, colchicine, and methotrexate, which result in malabsorption of nutrients (Roe 1985). Functional interactions also include drug effects on central and peripheral appetite-control mechanisms that affect food intake (Levitsky 1984).

OUTCOMES OF DIET–DRUG INTERACTIONS

Outcomes of diet and drug interactions can be clinically desirable or unwanted. Desirable diet and drug interactions include those that result in *control of disease processes* (Caranasos and Stewart 1985). These include prolongation of the effect of coumarin anticoagulants by dietary vitamin K restriction. An example of a diet–drug interaction leading to *loss of disease control* is the effect of a high-protein diet or formula on theophylline-dependent control of asthma (Feldman et al. 1980).

Acute toxic reactions include disulfiram reactions in alcoholics who drink while receiving the alcohol-aversion drug disulfiram (Antabuse); tyramine reactions due to ingestion of high-tyramine cheeses by those receiving monamine oxidase inhibitor drugs; and histamine reactions in those receiving isoniazid (Roe 1984; Uragoda 1980). Nutrients can be used to combat the adverse effects of a drug. For example, vitamin B_6 can be administered to combat the acute toxic effects of isoniazid (Brown 1972).

Growth retardation in children can result from use of amphetamines or other stimulant drugs, such as methyl phenidate, which reduces appetite. Nutritional stunting due to protein-losing enteropathy (excessive protein loss from the intestine) can be fought by drugs, including thiabendazole, used in the treatment of hookworm infestation. The growth spurt, however, which will follow effective treatment of hookworm infestation, is associated not only with the administration of the anthelmintic drug but also with increased food intake (Passmore and Eastwood 1986).

Nutritional deficiencies can result from food and drug interactions in the intestine, resulting in malabsorption of nutrients. An example is the malabsorption of dietary folate associated with the drug sulfasalazine (Swinson et al. 1981).

Nutritional support can result from the use of anti-infective or cancer chemotherapeutic drugs that bring about disease control. Thus, the patient who has had successful chemotherapy for Hodgkin's disease will gain back weight that was lost during the period of disease activity (Roe 1985b).

Blockage of nasogastric tubes used for enteral feeding may be due not only to administering pills via the tube, but also to giving liquid drug formulations that are incompatible with the formula constituents (Cutie et al. 1983).

DIAGNOSIS OF DRUG AND DIET INTERACTIONS

The following criteria, adapted from those proposed for general use with suspected drug reactions, can be applied to the diagnosis and differentiation of real from apparent adverse outcomes of drug and diet interactions (Folb 1980).

1 The association between intake of the drug, or drug and food combination, and the adverse outcome should be consistent.
2 There should be a reasonable temporal relationship between the intake of the drug, or drug and food combination, and the observed effect.
3 The association should be plausible.

However, as with other adverse drug reactions, the interaction with diet may be influenced by drug characteristics, by the drug regimen, or by characteristics of the individual taking the drug.

CLINICAL IMPORTANCE OF DRUG AND DIET INTERACTIONS

1. The Intended and Required Therapeutic Effect of the Drug Is Not Achieved. The intended and desired effect of a drug is not achieved as a result of drug–diet interaction if a food, beverage, or single nutrient reduces the amount of drug absorbed to subtherapeutic levels in the body. There may also be a resultant change in the rate of metabolism of the drug or accelerated clearance of the drug from the body such that its duration of action is insufficient to control the patient's symptoms.

2. Adverse Side Effects, Including Acute Incompatibility Reactions, Occur. Adverse side effects that occur as a result of drug and diet interactions include the following:

   a. Flush reactions including histamine reactions, disulfiram reactions, and chlorpropamide–alcohol flush reactions.
   b. Hypertensive attacks in patients on monamine oxidase inhibitor drugs, following ingestion of tyramine or dopamine-containing foods or beverages.
   c. Potentiation of drug effects, such as hemorrhage occurring in patients on coumarin anticoagulant drugs when they take megadoses of vitamin E.

3. Drug-induced Nutritional Deficiencies Are Produced. Drug-induced nutritional deficiencies may develop under the following circumstances:

   a. When the drug has a depressant effect on appetite and food intake. Protein-energy malnutrition (PEM) as well as micronutrient deficiencies can result from drug-induced anorexia.
   b. When the drug impairs digestion and causes nutrient malabsorption.
   c. When the drug inhibits or potentiates nutrient metabolism.
   d. When the drug potentiates nutrient losses from the body.

4. The Patient May Discontinue Use of the Drug When a Need Still Exists. The patient may discontinue the use of the drug while the need still exists either because the drug appears ineffective in alleviating symptoms or because of unpleasant side effects. Either of these outcomes can occur because of diet and drug interactions.

5. The Secondary Effect of a Drug–Diet Interaction May Be to Cause an Accident with Serious or Fatal Outcome. Automobile accidents may occur as a result of the potentiation of alcohol effects by a therapeutic drug. Even minor falls that occur in conjunction with drug-induced osteoporosis may involve serious bone fractures.

LIMITATIONS ON OUR KNOWLEDGE OF THE CLINICAL IMPORTANCE OF DRUG AND DIET INTERACTIONS

Despite the fact that single-dose studies have demonstrated that food or specific food components affect drug bioavailability, it does not necessarily follow that the therapeutic efficacy of the drug is compromised by the food effect under clinical conditions, when the drug is given at specific intervals (Melander 1984).

We also know that while drugs can have adverse effects on nutritional status, these can be offset by increased nutrient intake from the diet or from nutrient supplements. Thus, while antacids can reduce folate absorption, a study of clinic patients found the folate status of antacid takers to be better than that of the controls. The higher intake of folic acid by the antacid takers from fortified breakfast cereals and from vitamin pills could explain the observations (Bailey 1984). Thus, intergroup differences in diet may determine the clinical course of nutrient depletion.

DRUG REGIMENS LIKELY TO PRODUCE CLINICALLY SIGNIFICANT DRUG–NUTRIENT INTERACTIONS

Data from the U.S. National Prescription Audit for 1982 show that of the top 25 prescription drugs, 19 have prominent diet interactions of clinical significance. These are hydrochlorothiazide, combination oral contraceptives, propranolol, triamterene, potassium, diazepam, digoxin, furosemide, theophylline, penicillin V-K, atropine, methyldopa, cimetidine, phenobarbital, ibuprofen, tetracycline, amitryptilline, trimethoprim, and erythromycin (Baum et al. 1985). Typical conditions for drug interferences include the following:

1. *Drugs Taken with Food.* Drugs commonly taken with food include antacids and aspirin. *Examples:* (a) Aluminum hydroxide–containing antacids can cause phosphate depletion and secondary osteomalacia (Bloom and Flinchum 1960). (b) Aspirin can cause folate depletion and hyperexcretion of ascorbic acid (Alter et al. 1971; Sahud and Cohen 1971).

2. *Drugs Taken with Nutrient Supplements.* *Examples:* Phenytoin may be taken with folic acid supplements because of the known adverse effect of phenytoin on folate status. However, pharmacological doses of folic acid reduce phenytoin absorption (Richens 1976; Reynolds 1967).

3. Drugs Taken with Alcohol. *Examples:* Drugs taken with alcohol include those commonly taken with meals containing alcoholic beverages or immediately after dinner, drugs that are formulated in an alcohol vehicle, and drugs taken by alcoholics. Common drug–alcohol interactions include potentiation of sedative drug effects by alcohol, disulfiram and tyramine reactions, chlorpropamide–alcohol flush, and alcohol-induced malabsorption of nutrients (Roe 1979).

4. Drugs Used Purposefully in Order to Achieve a Specific Drug–Nutrient Interaction. *Examples:* Disulfiram is used as an alcohol-aversion drug. Coumarin anticoagulants, which are vitamin K antagonists, are used to prevent thromboses. Methotrexate, a folate antagonist, is effective as a cytotoxic drug by virtue of its antifolate effects (Morgan and Cagan 1974; Bjornsson 1984; Hackman and Hurley 1984).

5. Drugs Taken in Multiple Drug Regimens in Which More Than One Drug Produces an Adverse Effect Because of Drug and Diet Interactions. *Examples:* Isoniazid and rifampicin, which can both have an adverse effect on vitamin D status, are taken by .patients who have tuberculosis. Similarly, phenytoin and phenobarbital, which adversely affect folate status, are taken by patients with seizure disorders (Roe 1985a).

6. Drugs Which Cause Nutrient Depletion and Which are Taken for Long Periods of Time by Patients. The risk of nutritional deficiency is greatest in those patients whose diet is marginal or who have a disease that adversely affects nutritional status. *Examples:* Drugs that are taken for extended periods of time include anticonvulsants such as phenytoin, antituberculosis agents such as isoniazid, and loop diuretics such as furosemide, taken by patients with congestive heart failure (Roe 1984).

The elderly are at special risk for adverse outcomes for the following reasons:

1   Greater number of drugs prescribed
2   Multiple drug therapies
3   Prolonged usage of drugs
4   Self-medication
5   Noncompliance
6   Age-related changes in pharmacokinetics and pharmacodynamics
7   Age-related decline in renal function
8   Marginal diets
9   Mineral depletion due to diet or disease
10 Lack of exposure to sunlight

11 Megavitamin usage
12 Memory loss (J. Roy Coll Physicians 1984)

Alcoholics are at risk for adverse effects of drug and diet interactions because of:

1 Reactions to alcohol-aversion drugs
2 Consumption of alcoholic beverages while taking drugs which produce incompatibility reactions
3 Noncompliance with Rx
4 Incidence of malnutrition due to the effects of alcohol
5 Use of alcohol-containing over-the-counter drugs
6 Use and abuse of over-the-counter drugs
7 Use and abuse of sedatives and other drugs whose effect is potentiated by alcohol (Roe 1979)

Host-related risk factors for adverse outcome of drug and diet interactions include the following:

1 The patient is an alcoholic (excess risk is for drug–alcohol incompatibility reactions).
2 The patient is pregnant (drug-induced fetal malnutrition with associated teratogenic effect).
3 The patient is a laxative abuser (drug-induced hypokalemia with malabsorption).
4 The patient is an antacid abuser (phosphate depletion).
5 The patient is a chronic arthritic (reactions due to anti-inflammatory drugs such as aspirin or indomethacin, which induce gastrointestinal blood loss and anemia).
6 The patient is a hypertensive with depression who is receiving an antidepressant monamine oxidase inhibitor drug (tyramine reaction).
7 The patient has pulmonary tuberculosis (drug-induced vitamin deficiencies due to concurrent use of multiple antituberculosis agents that cause vitamin B_6, niacin, and functional vitamin D deficiency).
8 The patient is an epileptic (folate and vitamin D deficiency related to chronic intake of drugs such as phenytoin and phenobarbital which are used in seizure control).

Predictive factors related to drug regimen that increase the risk of drug and diet interactions include:

1 Adverse effect of the drug–diet interaction is related to the intended therapeutic effect of the drug.

2 Prescription instructions indicate that the drug should be taken with a food with which the drug interacts.
3 The patient is not told how, when, or when not to take the drug in relation to food or to a beverage.
4 The drug is intended for long-term use.
5 More than one drug is being given that imposes a risk of similar or additive adverse outcomes resulting from drug and diet interactions.

Multiple risk factors for adverse effects of drug and diet interactions may coexist in a patient and may collectively contribute to the adverse outcomes of drug and diet interactions, particularly drug-induced nutritional deficiencies. It is essential to understand that most of these adverse outcomes are preventable if the physician and the patient receiving the drug know the risks and how they may be avoided.

REFERENCES

ALTER, H.J., M.J. ZVAIFLER, and C.E. RATH. 1971. Interrelationship of rheumatoid arthritis, folic acid, and aspirin. *Blood* 38: 405–16.

ASATOOR, A.M., A.J. LEVI, and M.D. MILNE. 1963. Tranylcypromine and cheese. *Lancet* 2: 733–34.

BAILEY, A. 1984. The Effects of Antacid Consumption on Folate Status. M.S. Thesis, Cornell University, Ithaca, NY.

BAUM, C., D.L. KENNEDY, M.B. FORBES, and J.K. JONES. 1985. Drug use and expenditure in 1982. *J.A.M.A.* 253: 382–86.

BJORNSSON, T.D. 1984. Vitamin K and vitamin K antagonists. In *Drugs and Nutrients: The Interactive Effects*. ed. D.A. Roe and T.C. Campbell. 429–73. New York: Marcel Dekker.

BLOOM, W.L. and D. FLINCHUM. 1960. Osteomalacia with pseudofractures caused by the ingestion of aluminum hydroxide. *J.A.M.A.* 174: 1327–30.

BROWN, C.V. 1972. Acute isoniazid poisoning. *Am. Rev. Resp. Dis.* 105: 206–16.

CARANASOS, G.J. and R.B. STEWART. 1985. Clinically desirable drug interactions. *Ann. Rev. Pharmacol. Toxicol.* 25: 67–95.

CUTIE, A.J., E. ALTMAN, and L. LENKEL. 1983. Compatibility of enteral products with commonly employed drug additives. *J. Parent. Ent. Nutr.* 7: 186–91.

FELDMAN, C.H., V.E. HUTCHINSON, C.E. PIPPENGER, T.A. BLUMENFELD, B.R. FELDMAN, and W.J. DAVIS. 1980. Effect of dietary protein and carbohydrate on theophylline metabolism in children. *Pediatrics* 66: 956–62.

FOLB, P.I. 1980. *The Safety of Medicines: Evaluation and Prediction.* p. 88. Berlin: Springer-Verlag.

GIBALDI, M. 1977. *Biopharmaceutics and Clinical Pharmacokinetics.* 2nd ed. 27–65. Philadelphia: Lea & Febiger.

GUBLER, C.J. 1984. Thiamin. In *Handbook of Vitamins.* ed. L.J. Machlin. 245–97. New York: Marcel Dekker.

HACKMAN, R.M. and L.S. HURLEY. 1984. Drug-nutrient interactions in teratogenesis. In *Drugs and Nutrients: The Interactive Effects*. ed. D.A. Roe and T.C. Campbell. 299–329. New York: Marcel Dekker.

HOSTETTER, I.H. 1984. Progressive glomerular injury: Roles of dietary protein and compensatory hypertrophy. *Pharm. Rev.* 36: 101S–107S.

9

10   DIET AND DRUG INTERACTIONS

JOURNAL OF THE ROYAL COLLEGE OF PHYSICIANS. 1984. Medication for the Elderly. A Report of the Royal College of Physicians. Vol. 18: 7.

KAPPAS, A., K.E. ANDERSON, A.H. CONNEY, and A.P. ALVARES. 1976. Influence of dietary protein and carbohydrate on antipyrine and theophylline metabolism in man. *Clin. Pharmacol. Ther.* 20: 643–53.

LEVITSKY, D.A. 1984. Drugs, appetite and body weight. In *Drugs and Nutrients: The Interactive Effects.* ed. D.A. Roe and T.C. Campbell. 375–408. New York: Marcel Dekker.

MELANDER, A. 1984. Influence of food and different nutrients on drug bioavailability. *World Rev. Nutr. Diet.* 43: 34–44.

MORGAN, R. and E.J. CAGAN. 1974. Acute alcohol intoxication, the disulfiram reaction and methyl alcohol intoxication. In *The Biology of Alcoholism.* vol. 3. ed. B. Kissin and H. Begleiter. 163–89. New York: Plenum.

PASSMORE, R. and M.A. EASTWOOD. 1986. *Davidson and Passmore Human Nutrition and Dietetics.* 8th ed. 468–69. New York: Churchill Livingstone.

REYNOLDS, E.H. 1967. Effects of folic acid on the mental state and fit frequency of drug treated epileptic patients. *Lancet* 1: 1086–88.

RICHENS, A. 1976. *Drug Treatment of Epilepsy.* London: Henry Kimpton.

ROE, D.A. 1979. *Alcohol and the Diet.* 119–48, 208–22. Westport, CT: AVI.

———. 1980. Nutrient and drug interactions. *Nutr. Rev.* 42: 141–54.

———. 1984. Therapeutic significance of drug-nutrient interactions in the elderly. *Pharmacol. Rev.* 36: 1098–1218.

———. 1985a. *Drug-Induced Nutritional Deficiencies.* 2nd ed. Westport, CT: AVI.

———. 1985b. Pathological changes associated with drug-induced malnutrition. In *Nutritional Pathology: Pathobiochemistry of Dietary Imbalances.* ed. H. Sidransky. 357–79. New York: Marcel Dekker.

ROSENTHAL, W.S., N.F. ADHAM, R. LOPEZ, and J.M. COOPERMAN. 1973. Riboflavin deficiency in complicated chronic alcoholism. *Am. J. Clin. Nutr.* 26: 858–60.

SAHUD, M.A. AND R.J. COHEN. 1971. Effect of aspirin ingestion and ascorbic acid levels in rheumatoid arthritis. *Lancet* 1: 937–38.

SWINSON, C.M., J. PERRY, M. LUMB, and A.J. LEVI. 1981. Role of sulfasalazine in the aetiology of folate deficiency in ulcerative colitis. *Gut* 22: 456–61.

URAGODA, C.G. 1980. Histamine poisoning in tuberculous patients after ingestion of tuna fish. *Am. Rev. Resp. Dis.* 121: 157–59.

Chapter 2

EFFECTS OF FOOD, NUTRIENTS, AND NUTRITIONAL STATUS ON DRUG DISPOSITION

Food components can promote or retard the absorption and metabolism of nutrients and drugs. The extent of food's effect on the absorption or metabolism of drugs depends on the characteristics of the food, when the food is consumed, whether fluids are taken with the food, and on the characteristics of the individual consuming the food. The efficiency of the diet in meeting nutrient needs and the therapeutic efficacy of drugs are dependent on those interactions.

EFFECTS OF DRUGS, SPECIAL DIETS, AND DIETARY SUBSTANCES ON NUTRIENT AND DRUG BIOAVAILABILITY

The physicochemical properties of the nutritional or pharmacological substance and also the physiological or pathophysiological conditions pertaining within the gastrointestinal tract affect absorption. Whereas the amount of a drug absorbed is largely dependent on its dissolution characteristics and on the extent of drug metabolism within the gastrointestinal tract, the absorption of a nutrient depends on the efficiency of the digestive and specific absorptive process. Gastrointestinal disease can alter both drug and nutrient absorption, and the therapeutic modalities used in the management of gastrointestinal disease can do so as well.

Important determinants of the efficiency of nutrient and drug absorption are shown below.

1 Nutrient source: whether from food or supplement, and if from food, whether naturally occurring or added as enrichment.
2 Nutrient form: including chemical specie, particle size, and solubility.

3 Macronutrient composition of the diet: including type and percent of carbohydrate, fat, and protein.
4 Micronutrient composition of the diet: including vitamins, minerals, and trace elements consumed concurrently.
5 Dietary fiber: including type(s), form, and amount consumed concurrently.
6 Nonnutrient components of the diet: including phytates, tannins, etc.
7 Gastric acidity: including presence or absence of hypochlorhydria or achlorhydria.
8 Peptic digestion: including age, disease, or drug-related inhibition.
9 Gastric emptying time: including disease or drug-related change in the residence time of the food in the stomach.
10 Pancreatic function: including reduced pancreatic exocrine secretion.
11 Intestinal surface area: including the area of the absorptive surface of the small intestine.
12 Intestinal enzyme function: including activity or otherwise of disaccharidases.
13 Intestinal absorptive function: including the integrity or otherwise of active or carrier transport systems.
14 Hepatobiliary function: including bile acid synthesis, bile secretion, and gall bladder function.
15 Intestinal microflora including the presence or absence of bacterial overgrowth or "bacterial sterilization."
16 Parasites: including the presence or absence of helminthic parasites which compete for nutrients.
17 Intestinal motility: including the presence or absence of intestinal stasis or diarrhea.
18 Age: including whether very young or very old.
19 Physiological status: including whether pregnant.
20 Nutritional status: including whether a specific form of malnutrition is present, e.g., iron deficiency.
21 Environmental: including exposure to ultraviolet light which induces vitamin D synthesis in the skin.
22 Gastrointestinal disease: including presence or absence of enteropathies, inflammatory bowel disease, cystic fibrosis, pancreatitis, hepatitis, or cirrhosis.
23 Renal function: including renal synthesis of vitamin D metabolites and nutrient losses via the urine.
24 Cutaneous losses: including nutrient losses via desquamation or sweat and photodegradation of vitamins in the skin.

25 Inborn errors of metabolism: including inborn errors of nutrient transport.
26 Drugs: including use or abuse of therapeutic drugs and alcohol, which impair nutrient absorption.

The following factors influence the efficiency of drug absorption:

1 Lipid solubility of the drug
2 Dissociation constant of the drug
3 Form and formulation of the drug, including particle size, surface area of the particles, drug mixture, presence and type of inert additives, and coatings of solid formulations
4 Rate of disintegration and dissolution in the gastrointestinal fluid
5 Drug stability in gastrointestinal fluids
6 Gastric emptying time
7 pH at the absorptive surface
8 Temporal relationship of intake to intake of food
9 Concurrent intake of fluid
10 Complexation, including complexation of the drug with another drug or food substance in the gastrointestinal tract
11 Adsorption, including adsorption of the drug to another drug or a food component such as dietary fiber in the gastrointestinal tract
12 Disease present, including any disease which can influence drug absorption
13 Disease intervention by drug or special diet that can influence drug adsorption (Anderson 1983; Forbes and Erdman 1983; Gibaldi 1977; Lynch and Cook 1980; Melander 1978; Parsons 1977; Roe 1984)

In order to understand how special diets or dietary components affect nutrient and drug absorption, it is important to understand the following concepts:

1 The specific macronutrient in a special diet can promote, retard, or reduce the absorption of drugs as well as nutrients.
2 Gastrointestinal disease can impair or promote drug absorption.
3 Special diets that are modified with respect to macronutrient composition may be employed in the management of gastrointestinal as well as metabolic disease.
4 The therapeutic effect of such a special diet may reverse the effects of gastrointestinal disease on drug absorption.

Diabetic autonomic neuropathy causes gastric stasis and thus alters

TABLE 2.1 EFFECTS OF DISEASE-RELATED CHANGES IN
GASTROINTESTINAL FUNCTION ON THE ABSORPTION
OF DRUGS AND NUTRIENTS

Changes in GI Function	Effects on Absorption of	
	Nutrients	Drugs
Gastric stasis	Slowed	Slowed
Gastrectomy	Reduced absorption Vitamin B_{12} Vitamin D Calcium & iron	Accelerated
Achlorhydria	Reduced iron absorption	Enhanced penicillin absorption. Reduced absorption. Diazepam derivative. (Chlorazepate [Grahnen 1985])
Proximal intestinal malabsorption syndrome	Reduced absorption fat, fat-soluble vitamins, folate, iron, & calcium	
Distal intestinal with rapid gut transit	Reduced absorption vitamin B_{12}	Reduced absorption sulfasalazine (Das et al. 1973)

the rate and efficiency of nutrient and drug absorption. Pyloric stenosis, cystic fibrosis, chronic pancreatitis, gluten-sensitive enteropathy (celiac disease), Crohn's disease, gastrectomy, and intestinal resection induce similar effects by slowing gastric emptying time, reducing availability of pancreatic exocrine secretion, and by reducing the intestinal absorptive surface and intestinal transit time (Parsons 1977).

Relationships between gastrointestinal disease and absorption capacity, relative to drugs and nutrients, is shown in Table 2.1.

Drug treatment of gastrointestinal disease can improve nutrient absorption. For example, when the H2 receptor antagonist cimetidine is administered to patients with gastric hypersecretion after small intestinal resection, nutrient absorption becomes more efficient with reduced fecal losses of fat and nitrogen (Cortot et al. 1979; Murphy et

TABLE 2.2 SPECIAL DIETS AND DIETARY SUBSTANCES
AFFECTING NUTRIENT AND DRUG ABSORPTION

Diet/Dietary Substance	Effect
High fiber (containing pectin, guar gum, or acarbose)	Delays sugar absorption
High-fiber diet (containing coarse bran)	Enhances riboflavin absorption
High-fiber diet crackers (containing psyllium)	Reduces riboflavin absorption
High-fiber diet (containing bran)	Reduces absorption of iron
Hypoallergenic diet (containing soy)	Reduces absorption of iron
Lactose-free diet (in lactose-intolerant patients)	Increases absorption of aspirin
Peptic ulcer diet (containing milk)	Decreases absorption of tetracycline

al. 1979). The effects of cimetidine are to inhibit gastric acid secretion, to reduce gastric fluid secretion, and to reduce the fractional gastric emptying rate. Thus, both maldigestion and malabsorption due to the late effects of GI surgery are lessened.

Special diets and dietary substances which alter nutrient and drug bioavailability are summarized in Table 2.2 (Berchtold et al. 1984).

Special diets that affect nutrient and drug absorption help us to identify the common mechanisms responsible for effects of food and beverages on bioavailability. For example, addition of fiber to the diet will slow gastric emptying time with one or more of the following effects:

A. *Increased solubility.* A drug or solid preparation of a vitamin has more time to dissolve in the gastric fluids prior to ejection of the nutrient or drug into the small intestine, where its absorption depends on it being in a soluble state. A model for demonstration of this effect is provided when riboflavin tablets are given with a glucose polymer solution (Polycose, Ross). Riboflavin absorption is then slowed relative to when the same tablets are given with milk. This is explained by the differential solubility of the solid formulation of the vitamin in these two beverages.

B. *More gradual uptake of glucose* from the proximal small intestine with resultant improvement in glucose tolerance (an important effect in the diabetic).

C. *More efficient delivery* of nutrients and drugs to a saturable absorption site in the proximal small intestine. It is known that most dietary substances which slow gastric emptying time enhance the absorption of riboflavin. Sodium alginate used as a food additive, cola beverages containing phosphoric acid and sugar, as well as coarse bran in breads and breakfast cereal all delay gastric emptying time and all enhance riboflavin absorption. A similar enhancement of riboflavin absorption can be produced by ingestion of large meals, hot meals, and meals containing fat because these slow gastric emptying time. Similarly, anticholinergic drugs that slow gastric emptying time enhance riboflavin availability (Roe 1984; Belko et al. 1982; Levy and Rao 1972; Brian Houston and Levy 1975; Roe et al. 1978; Jusko and Levy 1975; Levy and Jusko 1966).

It may seem paradoxical to note that increasing nutrient absorption in this way may entail a lower rate of absorption. Drugs that are slow to dissolve and/or are better absorbed when they reach a saturable absorption site at a slower rate are more available when they are given with food substances or conditions which promote the absorption of riboflavin. The improved absorption of the diuretic chlorothiazide is explained in this manner (Osman et al. 1982).

Conversely, special diets and dietary substances as well as components of normal foods can inhibit nutrient and drug absorption if they interact with one another in the gut lumen. Products containing psyllium seed reduce the absorption of riboflavin. A similar effect may explain the observation that in the rat, a high-fiber diet containing more than 10% of a mixture of methylcellulose, agar, wheat grain, and pectin decreases the absorption of beta-carotene. Interestingly, a similar fiber source added to the rat's diet at a level of 5% of the diet promotes beta-carotene absorption, perhaps because of effects on stomach emptying (Gronowska-Senger et al. 1980).

The effects of dietary fiber on iron bioavailability are apparently highly dependent on the ligands present in the meal. Several studies have indicated that absorption of nonheme iron as in fortified food or iron supplements is reduced by inclusion of fiber in the diet (McCance and Widdowson 1942; Jenkins et al. 1975; Mameesh et al. 1970; Dobbs and McLean-Baird 1977; Bjorn-Rasmussen 1974). It has, in fact, been proposed that iron nutriture may be impaired by consumption of high-fiber diets (Cummings 1978; Institute of Food Technologists 1979).

However, findings from experiments carried out by Leigh and Miller (1983) indicate that diets rich in meats and vegetables may produce better absorption of nonheme iron than carbohydrate diets with similar fiber contents because of the promoting effects of ascorbic acid present in the vegetables and of small-molecular-weight ligands derived from the meat. Nevertheless, co-administration of fiber sources such as bran with therapeutic iron preparations may reduce iron absorption.

It has been shown that the reduction in iron absorption by wheat bran may be brought about by soluble, phosphate-rich, nonphytate-containing components of the bran (Simpson et al. 1981). The low availability of iron from milk products has been explained by the presence of phosphate (Mahoney and Hendricks 1978).

Although a number of nonnutrient food and beverage components may reduce iron absorption, it has been pointed out by Forbes and Erdman (1983) that relatively few have practical importance. Tannins in tea certainly impair iron absorption and this may be very important among heavy tea drinkers on diets that are low in iron (Disler et al. 1975). While strong chelating agents such as disodium EDTA, which is used as an intentional food additive, reduces the availability of iron, its practical importance is questioned because the dietary occurrence of EDTA is very low. It is of interest that ferric EDTA has been used as an iron fortification compound (Cook et al. 1981).

Soy protein products inhibit nonheme iron absorption. However, addition of meat or 100 mg of ascorbic acid to a meal containing isolated soy protein can increase iron availability (Cook et al. 1981; Morck et al. 1982).

The bioavailability of zinc from high-fiber diets, particularly vegetarian diets, is low. It is rather generally accepted that phytic acid is the inhibitor of zinc absorption in foods of plant origin (Forbes and Erdman 1983).

It is important to consider the therapeutic or positive effects of special diets on nutrient and drug absorption. Thus while lactose intolerance reduces the absorption of aspirin, a low-lactose diet has the reverse effect. Similarly, maintenance of celiac disease patients on a gluten-free diet normalizes folate absorption (Floch 1981).

FOOD EFFECTS ON DRUG BIOAVAILABILITY

Food can influence drug bioavailability because of physicochemical or biochemical interactions between a specific nutrient or other food component and drug molecules within the gastrointestinal tract. Moreover, it has been amply demonstrated that gastrointestinal events including change in motility, secretion, or microflora that are due to the diet can explain the observed food effects (Melander 1978; Welling 1977; Toothaker and Welling 1980).

In considering effects of food on drug bioavailability, it is convenient to group drugs according to their therapeutic indication and usage. However, increasingly, it is becoming clear that for any one drug, different food components can alter absorption in opposing directions or to a differing extent in the same direction. Thus one may account for the variation in findings between investigators who have used "standard breakfasts" of different composition to test effects of food on drug bioavailability.

Examples of apparent contradiction in the effect of food on drug absorption may be explained as in the case of phenytoin, by differences in the composition of the meal fed with the drug. For example, it was shown by Melander et al. (1979) that a micronized preparation of phenytoin was better absorbed after a standard Swedish breakfast than when the same formulation of the drug was given to the same subjects in the fasting state.

When the same drug formulation was given with an amount of carbohydrate equivalent to that in the original breakfast test meal (50 g), a similar enhancement of phenytoin absorption was obtained. On the other hand, when the same preparation of phenytoin was given with a food source of protein, there was a reduction in the amount of phenytoin absorbed. The protein source used in this study consisted of concentrated cow's milk proteins (Johansson et al. 1983). The investigators suggest that this protein source could have impaired phenytoin absorption because the albumin present became bound to the drug and the protein–drug complex was then retained in the gut. Another possibility offered is that the protein acts as a buffer at a pH outside the optimal range for absorption of the phenytoin. It has been shown by in vitro studies that while protein may have a significant effect on the dissolution of phenytoin, the effect depends on the pH, the nature of the protein, and the protein concentration (Rosen and Macheras 1984).

Phenytoin is also less well absorbed when there is concurrent administration of folic acid (Bayless et al. 1971). It may therefore be anticipated that the same effect might be obtained if the drug were given with food sources of this vitamin including particularly large portions of folic acid–fortified breakfast cereals.

Co-administration of phenytoin with high doses of another B vitamin, pyridoxine, may also reduce the absorption of the drug (Hansson and Silenpaa 1976).

Different methods of assay for drugs and their metabolites may explain some variability in the results of drug absorption studies. Thus, while Thulin et al. (1983) found that the absorption of spironolactone was similar whether it was given in the fasting state or after a standard Swedish breakfast, these same investigators had previously shown that this type of breakfast enhanced the absorption of canrenone, a metabolite of spironolactone after a single dose of the parent drug had been given (Melander et al. 1977).

Recognizing the effect of food on drug absorption may lead to more precise information on appropriate dosages and intervals for drugs that are relatively insoluble or appear to have an erratic absorption pattern. For example, the antifungal agent griseofulvin had been recommended for children in a variety of dosage schedules. Food interference studies then indicated that the drug was better and more predictably absorbed after ingestion of whole milk (Tulpule and Krishnaswamy 1983). This observation concurs with observations that griseofulvin is better absorbed when given after a fatty meal (Gerard et al. 1984).

We have further observed that in patients on self-prescribed or physician-prescribed low-fat diets, the effectiveness of griseofulvin is significantly reduced (personal observation).

BEVERAGE EFFECTS ON DRUG DISPOSITION

FLUID INTAKE

Drug absorption is affected by fluid intake. The rate of drug absorption may be decreased by intake of large volumes of fluids such as soft drinks. Colas containing phosphate and sugar may decrease the rate of drug absorption because they delay the rate of stomach emptying. However, the efficiency of drug absorption may be increased by concurrent intake of beverages. Reasons include speedier dissolution of the drug, osmotic effects, or change in the surface area of the intestinal mucosa to which the drug molecules are exposed. Further, the temperature of the beverage as well as its composition and volume may affect drug absorption (Welling 1977; Borowitz et al. 1971; Welling 1980).

MILK AND MILK FORMULAS

Ingestion of milk has long been known to interfere with the absorption of tetracyclines because insoluble chelates are formed between the drug and the calcium present in the beverage (Neuvonen et al. 1971). It has also been shown that milk decreases the bioavailability of the beta-adrenergic blocking agent sotalol. It has been suggested that this effect may also be due to an interaction with calcium (Kahela et al. 1979).

In premature infants, a milk feed does not influence the total amount of theophylline absorbed, but decreases the rate of absorption, probably because of effects of the milk on stomach emptying (Heimann et al. 1982).

CAFFEINE-CONTAINING BEVERAGES

Caffeine is consumed not only in coffee but also in other beverages, including tea, colas, and other sodas. Monks et al. showed that elimination of methylxanthines, such as caffeine, from the diet led to faster and more extensive metabolism of theophylline (Monks et al. 1979). However, addition of extra methylxanthines to the diet did not significantly alter the disposition of theophylline. It has been suggested that even if there were a significant effect of caffeine on theophylline disposition, it may not be therapeutically important because most people do not materially change the caffeine content of their diet in an abrupt manner (Jonkman and Upton 1984). Our observations of clinic or office patients, however, find many who may change the caffeine content of their diets in the pursuit of general health as well as for the management of peptic ulcer, gastritis, rosacea, and headaches.

FOOD SUBSTANCE, NUTRIENT, AND NUTRITIONAL EFFECTS ON DRUG METABOLISM

Food substances responsible for variation in metabolism of and responsiveness to drugs may be grouped, based on recent review of the literature as follows (Alvares 1984):

1 Macronutrients (carbohydrate, protein, fat)
2 Trace substances (vitamins, trace elements)
3 Lipid-soluble xenobiotics (including constituents of spices, e.g., safrole, herbs, smoked food, broiled meats, vegetables, and fruits).

Nonnutrient food substances and nutrients that may affect drug metabolism are shown in Table 2.3, together with mechanisms responsible for these effects.

Unfortunately, however, in the present state of knowledge, the significance of the diet in modifying drug metabolism cannot be precisely estimated. Indeed, it is necessary to consider that variables of age, sex, pharmacogenetic inheritance, smoking, alcohol intake, occupation, environment, and tissue function also have large effects on drug disposition (Vesell 1984).

EFFECTS OF MACRONUTRIENTS ON DRUG DISPOSITION

Major changes in the macronutrient composition of the diet can have significant effects on the rate of metabolism of therapeutic drugs. Perhaps the most important effect results from a change in protein intake. High-protein diets increase, while low-protein diets decrease the rate

TABLE 2.3 EFFECT OF NUTRIENTS, NONNUTRIENTS IN FOOD,
AND FOOD PREPARATION ON DRUG DISPOSITION

Substance or Process	Effect
Protein (high-protein diet)	Enhances rate of theophylline metabolism
Indoles and flavonoids	Enhance rate of drug metabolism
Alcohol	High intake inhibits, and chronic abuse promotes drug metabolism
Charcoal broiling	Enhances rate of metabolism

Sources: Kappas et al. 1976; Alvares 1984; Vessell et al. 1971; Pantuck 1976.

of drug metabolism. This has been demonstrated for certain model drugs, including theophylline, but may apply to other drugs in therapeutic use. Consequent changes in the blood levels of the drug may occur when a patient goes on or off a special diet or changes the nutrient formula consumed. If a change is made from a high- to a low-protein diet, longer retention of the drug could present higher risk of a toxic reaction. Raising the protein content of the diet may eliminate the desired therapeutic effects. In pediatric patients with asthma, it has been observed that wheezing reappeared more rapidly when theophylline was administered with a high-protein compared to a low-protein diet (Anderson et al. 1982).

EFFECTS OF NONNUTRIENT FOOD SUBSTANCES

An extensive group of nonnutrient, lipid-soluble substances present in food can alter the rate of drug metabolism. These substances include naturally occurring food constituents, substances formed in the food as a result of processing or cooking, food additives, and contaminants. The effect of a nonnutrient food substance depends not only on the interaction with therapeutic drugs but also on the nutritional status of the person consuming the food chemical: frequency of consumption, the microflora of the intestinal tract, and environmental exposures to chemicals, light, or ionizing radiation.

Adverse acute outcomes of interactions may be dose-dependent and are certainly related to the temporal pattern of intake in relation to the interacting drug. Chronic adverse outcomes, including mutagenesis and carcinogenesis, may be influenced by a patient's sex, age, pharmacogenetic characteristics, smoking and alcohol habits, nutritional status, and hepatic and renal function (Alvares et al. 1976; Alvares et al. 1979; Mucklow et al. 1982; Wattenberg 1971).

TABLE 2.4 REPORTING EFFECTS OF PROTEIN-ENERGY
MALNUTRITION ON DRUG DISPOSITION

Drug Disposition	Changes in Infants	Changes in Chronic PEM in Adults
Drug absorption		Tetracycline absorption reduced
Protein binding	Decrease in binding of cloxacillin, streptomycin, sulphamethoxazole, digoxin, thiopentone, warfarin	Decrease in binding of sulfadiazine, tetracycline, phenylbutasone
Volume of distribution	Gentamicin, cefoxitin, isoniazid	Tetracycline
Metabolism	Acetylation, INH slowed	Acetylation, sulfadiazine faster
Clearance	Clearance of tetracycline faster	Clearance of penicillin, gentamicin, tobramycin, cefoxitin slowed

Sources: Buchanan 1984; Ragmuram and Krishnaswamy 1981; Cusack and Denham 1984.

Since the effects of food on drug absorption are complex, it is useful to provide simple guidelines which emphasize the proper timing of drug dosage:

1 Take the medication daily at the same time in relation to eating or consuming nourishing beverages or food supplements.
2 Do not consume any specific foods or beverages (including alcohol) which can cause reactions to the medication.
3 Report to your physician any signs of gastric irritation or other unpleasant side effect which you experience on taking the drug.
4 Check with your physician if the drug appears to have an erratic effect.
5 Check with your pharmacist to be sure you are following the safest and best dosage schedule for all your prescription medications.
6 Do not start taking nutrient supplements while on prescription medicines unless approved by your physician or pharmacist.

EFFECT OF CHANGE IN NUTRITIONAL STATUS ON THE RATE OF DRUG METABOLISM

Food and protein restriction may result in protein-energy malnutrition (PEM), but in any case can affect the rate of drug metabolism and also the rate of drug elimination in humans and laboratory animals (Krishnaswamy et al. 1984).

For drugs that are highly protein bound, the rate of elimination may be increased by PEM or hypoalbuminemia. However, drugs that are not highly protein bound may be eliminated more slowly because of reduced renal function occuring with severe PEM (Buchanan 1984).

Reported effects of protein-energy malnutrition on the disposition of specific drugs are summarized in Table 2.4.

REFERENCES

ALVARES, A.P. 1984. Environmental influences on drug biotransformations in humans. In *Nutrition, Food and Drug Interactions in Man.* ed. G. Debry. 45–59. *World Rev. Nutr. Dietet.* vol. 43. Basel: Karger.

ALVARES, A.P., K.E. ANDERSON, A.H. CONNEY, and A. KAPPAS. 1976. Interactions between nutritional factors and drug biotransformations in man. *Proc. Natl. Acad. Sci. USA* 73: 2501–4.

ALVARES, A.P., A. KAPPAS, J.L. EISEMAN, K.E. ANDERSON, C.B. PANTUCK, E.J. PANTUCK, K-C. HSAIO, W.A. GARLAND, and A.H. CONNEY. 1979. Intraindividual variation in drug disposition. *Clin. Pharm. Ther.* 26: 407–19.

ANDERSON, C.E. 1983. Vitamins. In *Nutritional Support of Medical Practice,* 2nd ed. 23–53. eds. H.A. Schneider, C.E. Anderson, and D.B. Coursin. Philadelphia: Harper and Row.

ANDERSON, K.E., A.H. CONNEY, and A. KAPPAS. 1982. Nutritional influences on chemical biotransformations in humans. *Nutr. Rev.* 40: 161–70.

BAYLESS, E.M., J.M. CROWLEY, J.M. PREECE, P.E. SYLVESTER, and V. MARKS. 1971. Influence of folic acid on blood phenytoin levels. *Lancet* 1: 62–64.

BELKO, A., M. ROTTER, and D.A. ROE. 1982. Glucose polymer effects on riboflavin and folic acid absorption. Abstract. *J. Am. Coll. Nutr.* 1: 413.

BERCHTOLD, P., T.R. WEIHRAUCH, and M. BERGER. 1984. Food and drug interactions on digestive absorption. In *Nutrition, Food and Drug Interactions in Man.* ed. G. Debry. 10–33. *World Rev. Nutr. Diet* vol. 43. Basel: Karger.

BJORN-RASMUSSEN, E. 1974. Iron absorption from wheat bread: Influence of various amounts of bran. *Nutr. Metab.* 16: 101–10.

BOROWITZ, J.L., P.F. MOORE, G.K.W. YIM, and T.S. MIYA. 1971. Mechanisms of enhanced drug effects produced by dilution of the oral dose. *Toxicol. Appl. Pharmacol.* 19: 164–68.

BRIAN HOUSTON, J. and G. LEVY. 1975. Effect of carbonated beverages and of an antiemetic containing carbohydrate and phosphoric acid on riboflavin bioavailability and salicylamide biotransformation in humans. *J. Pharm. Sci.* 64: 1504–7.

BUCHANAN, N. 1984. Effect of protein-energy malnutrition on drug metabolism in man. *World Rev. Nutr. Diet* 43: 129–39. Basel: Karger.

COOK, J.D., T.A. MORCK, and S.R. LYNCH. 1981. The inhibitory effect of soy products on non-heme iron absorption in man. *Am. J. Clin. Nutr.* 34: 2622–29.

CORTOT, A., C.R. FLEMING, and J-R. MALAGELEDA. 1979. Improved nutrient absorption after cimetidine in short-bowel syndrome with gastric hypersecretion. *New Eng. J. Med.* 300: 79–80.

CUMMINGS, J.H. 1978. Nutritional implications of dietary fiber. *Am. J. Clin. Nutr.* 31: 521–29.

CUSACK, B. and M.J. DENHAM. 1984. Nutritional status and drug disposition in the elderly. In *Drugs and Nutrition in the Geriatric Patient.* ed. D.A. Roe. 71–91. New York: Churchill Livingstone.

DAS, K.M., M.A. EASTWOOD, P.A. McMANUS, and W. SIRCUS. 1973. The metabolism of salicylazoslfapyridine in ulcerative colitis. *Gut* 14: 631–41.

DISLER, P.B., S.R. LYNCH, J.D. TORRENCE, M.H. SAYERS, T.H. BOTHWELL, and R.W. CHARLTON. 1975. The mechanism of the inhibition of iron absorption by tea. *S. Afr. J. Med. Sci.* 40: 109–16.

DOBBS, R.T. and J. McLEAN-BAIRD. 1977. Effect of whole meal and white bread on iron absorption in normal people. *Brit. Med. J.* 25: 1641–42.

FLOCH, M.H. 1981. *Nutrition and Diet Therapy in Gastrointestinal Disease.* 249–64. New York: Plenum.

FORBES, R.M. and J.W. ERDMAN. 1983. Bioavailability of trace mineral elements. *Ann. Rev. Nutr.* 3: 213–31.

GERARD, J., P.J. LEFEBRE, and A.S. LUYCKX. 1984. Glibenclamide pharmacokinetics in carbose treated type 2 diabetics. *Europ. J. Clin. Pharmacol.* 27: 233–36.

GIBALDI, M. 1977. *Biopharmaceutics and Clinical Pharmacokinetics.* 2nd ed. 15–66. Philadelphia: Lea and Febiger.

GRAHNEN, A. 1985. A discussion of drug metabolism in the aged. *Drug-Nutrient Interact.* 4: 107–13.

GRONOWSKA-SENGER, A., E. SMACZNY, and B. KOBKOWICZ. 1980. The effect of fiber on the utilization of carotene in laboratory rats. *Bromatol. Chemia Toksykologiczna* 13: 129–34.

HANSSON, O. and M. SILENPAA. 1976. Pyridoxine and serum concentrations of phenytoin and phenobarbitone. *Lancet* 1: 256.

HEIMANN, G., J. MURGESCU, and U. BERGT. 1982. Influence of food intake on bioavailability of theophylline in premature infants. *Europ. J. Clin. Pharmacol.* 22: 171–73.

HOLLOWAY, E.D. and F.J. PETERSON. 1984. Ascorbic acid in drug metabolism. In *Drugs and Nutrients: The Interactive Effects.* ed. D.A. Roe and T.C. Campbell. 225–95. New York: Marcel Dekker.

INSTITUTE OF FOOD TECHNOLOGISTS. Expert Panel on Food Safety and Nutrition and the Committee on Public Information. 1979. Dietary fiber. *Food Technol.* 33: 35–39.

JENKINS, D.J.A., M.A. HILL, and J.H. CUMMINGS. 1975. Effect of wheat fiber on blood lipids, fecal steroid excretion and serum iron. *Am. J. Clin. Nutr.* 28: 1408–11.

JOHANSSON, O., E. WAHLIN-BOLL, T. LINDBERG, and A. MELANDER. 1983. Opposite effects of carbohydrate and protein on phenytoin absorption in man. *Drug-Nutrient Interactions* 2: 139–44.

JONKMAN, J.H.G. and R.A. UPTON. 1984. Pharmacokinetic drug interactions with theophylline. *Clin. Pharmacokinet.* 9: 309–34.

JUSKO, W.J. and G. LEVY. 1975. Absorption, protein binding and elimination of riboflavin. In *Riboflavin.* ed. R. Rivlin. 99–151. New York: Plenum.

KAHELA, P., M. ANTILA, R. TIKKANEN, and H. SUNDQUIST. 1979. Effect of food, food constituents, and fluid volume on the bioavailability of sotalol. *Acta Pharmacol. Toxicol.* 44: 7–12.

KAPPAS, A., A.P. ALVARES, K.E. ANDERSON, E.J. PANTUCK, C.B. PANTUCK, R. CHANG, and A.H. CONNEY. 1978. Effects of charcoal broiled beef on antipyrine and theophylline metabolism. *Clin. Pharmacol. Ther.* 23: 445–50.

KAPPAS, A., K.E. ANDERSON, A.H. CONNEY, and A.P. ALVARES. 1976. Influence of dietary protein and carbohydrate on antipyrine and theophylline metabolism in man. *Clin. Pharmacol. Ther.* 20: 643–53.

KRISHNASWAMY, K., R. KALAMEGHAM, and N.A. NAIDU. 1984. Dietary influences on the kinetics of antipyrine and aminopyrine in human subjects. *Br. J. Clin. Pharmac.* 17: 139–46.

LEIGH, M.J. and D.D. MILLER. 1983. Effects of pH and chelating agents on iron binding by dietary fiber: implications for iron availability. *Am. J. Clin. Nutr.* 38: 202–13.

LEVY, G. and W.J. JUSKO. 1966. Factors affecting the absorption of riboflavin in man. *J. Pharm. Sci.* 55: 285–89.

LEVY, G. and B.K. RAO. 1972. Enhanced intestinal absorption of riboflavin from sodium alginate solution in man. *J. Pharm. Sci.* 61: 279–80.

LYNCH, S.R. and J.D. COOK. 1980. Interaction of vitamin C and iron in micronutrient interactions: vitamins, minerals and hazardous elements. *Ann. N.Y. Acad. Sci.* 355: 32–44.

MAHONEY, A.W. and D.G. HENDRICKS. 1978. Some effects of different phosphate compounds on iron and calcium absorption. *J. Food Sci.* 43: 1473–76.

MAMEESH, M., S. APRANHAMIAN, J.P. SALJC, and J.W. COWAN. 1970. Availability of iron from labelled wheat, chickpea, broad bean and okra in anemic blood donors. *Am. J. Clin. Nutr.* 23: 1027–32.

McCANCE, R.A. and E.M. WIDDOWSON. 1942. Mineral metabolism of healthy adults on white and brown bread dietaries. *J. Physiol.* 101: 44–85.

MELANDER, A. 1978. Influence of food on the bioavailability of drugs. *Clin. Pharmacokinet.* 3: 337–51.

MELANDER, A., G. BRANTE, O. JOHANNSSON, T. LINDBERG, and T. WAHLIN-BOLL. 1979. Influence of food on the absorption of phenytoin in man. *Eur. J. Clin. Pharmacol.* 15: 269–74.

MELANDER, A., K. DANIELSON, B. SCHERSTEN, T. THULIN, and E. WAHLIN. 1977. Enhancement by food of canrenone bioavailability from spironolactone. *Clin. Pharmacol. Ther.* 22: 100–103.

MONKS, T.J., J. CALDWELL, and R.L. SMITH. 1979. Influence of methylxanthine-containing foods on theophylline metabolism and kinetics. *Clin. Pharmacol. Ther.* 26: 513–24.

MORCK, T.A., S.R. LYNCH, and J.D. COOK. 1982. Reduction of the soy-induced inhibition of non-heme iron absorption. *Am. J. Clin. Nutr.* 36: 319–28.

MUCKLOW, J.C., M.T. CARAHER, D.B. HENDERSON, P.H. CHAPMAN, D.F. ROBERTS, and M.D. RAWLINS. 1982. The relationship between individual dietary constituents and antipyrine metabolism in Indo-Pakistani immigrants to Britain. *Brit. J. Clin. Pharmacol.* 13: 481–86.

MURPHY, J.P., D.R. KING, and A. DUBOIS. 1979. Treatment of gastric hypersecretion with cimetidine in the short-bowel syndrome. *New Eng. J. Med.* 300: 80–81.

NEUVONEN, P., M. MATILLA, G. GOTHINI, and R. HACKMAN. 1971. Interference of iron and milk with absorption of tetracycline. *Scand. J. Clin. Lab. Invest.* 27: 76.

OSMAN, M.A., R.B. PATEL, D.S. IRWIN, W.A. CRAIG, and P.G. WELLING. 1982. Bioavailability of chlorothiazide from 50, 100 and 250 mg solution doses. *Biopharmaceut. Drug Dispos.* 3: 89–94.

PANTUCK, E.J., K-C. HSIAO, W.D. LOUB, L.W. WATTENBERG, R. KUNTZMAN, and A.H. CONNEY. 1976. Stimulatory effect of vegetables on intestinal drug metabolism in the rat. *J. Pharm. Exp. Ther.* 198: 278–83.

PARSONS, R.L. 1977. Drug absorption in gastrointestinal disease with particular reference to malabsorption syndromes. *Clin. Pharmacokinet.* 2: 45–60.

RAGMURAM, T.C. and K. KRISHNASWAMY. 1981. *Drug-Nutrient Interact.* 1: 23–29.

ROE, D.A. 1984. Food, formula and drug effects on the disposition of nutrients. In *Nutrition, Food and Drug Interactions in Man.* ed. G.S. Debry. 80–94. *World Review of Nutrition and Dietetics,* 43. Basel: Karger.

ROE, D.A., K. WRICK, D. McLAIN, and P. VAN SOEST. 1978. Effects of dietary fiber sources on riboflavin absorption. *Fed. Proc.* 37: 756.

ROSEN, A. and P. MACHERAS. 1969. The effect of protein on the dissolution of phenytoin. *J. Pharm. Pharmacol.* 36: 723–27.

SAPEIKA, N. 1969. *Food Pharmacology.* 139–44. Springfield, IL: Charles C. Thomas.

SIMPSON, K.M., E.R. MORRIS, and J.D. COOK. 1981. The inhibitory effect of bran on iron absorption in man. *Am. J. Clin. Nutr.* 34: 1469–78.

THULIN, T., E. WAHLIN-BOLL, L. LIEDHOLM, and A. MELANDER. 1983. Influence of food intake on antihypertensive drugs: spironolactone. *Drug-Nutrient Interactions* 2: 169–73.

TULPULE, A. and K. KRISHNASWAMY. 1983. Effect of rice diet on chloroquine bioavailability. *Drug-Nutrient Interactions* 2: 83–86.

TOOTHAKER, R.D. and P.G. WELLING. 1980. The effect of food on drug bioavailability. *Ann. Rev. Pharmacol. Toxicol.* 20: 173–79.

VESELL, E.S. 1984. Complex effects of diet on drug disposition. *Clin. Pharmacol. Ther.* 36: 285–95.

VESELL, E.S., J.G. PAGE, and G.T. PASSANANTI. 1971. Genetic and environmental factors affecting ethanol metabolism in man. *Clin. Pharmacol. Ther.* 12: 192–201.

WATTENBERG, L.W. 1971. Studies of polycyclichydrocarbon hydroxylases of the intestine possibly related to cancer. Effect of diet on benzpyrene hydroxylase activity. *Cancer* 28: 99–102.

WELLING, P.G. 1977. Influence of food and diet on gastrointestinal drug absorption: a review. *J. Pharmacokinet. Biopharm.* 5: 291–334.

_____. 1980. The effect of food on drug bioavailability. *Ann. Rev. Pharmacol. Toxicol.* 20: 173–99.

Chapter 3

NUTRIENT AND DRUG–NUTRIENT
INTERACTIONS DUE TO FORMULA FOODS

Formula foods are used to supply liquid diets to infants and to individuals of all ages whose capacity to ingest, absorb, or utilize conventional food sources is markedly impaired. Formula foods may be prepared from milk, cream, eggs, and other common foods, but are usually semisynthetic. The nitrogen source within the formulas may be milk powder, casein, hydrolyzed casein, soy protein isolate, or a mixture of essential and nonessential amino acids. The carbohydrate in these diets consists of corn syrup solids, modified tapioca starch, sucrose, glucose, and lactose as well as glucose polymers. Fat in the formula may be hydrogenated soy oil, coconut oil, or medium chain triglycerides. Fat- and water-soluble vitamins, minerals, and trace elements are added at levels to supply or slightly exceed daily requirements. Intentional additives are added to formulas as antioxidants or stabilizers. Formulas are prepared for use in oral feeding and for use in enteral or parenteral hyperalimentation. They may be used to supplement the diet or to supply total daily needs for food-energy and nutrients.

Issues of concern with respect to formula foods and feedings are:

1 Composition of the mixture (including both nutrients and toxicants).
2 Stability of the mixture.
3 Compatibility of components in the mixture.
4 Allergenicity of the ingredients.
5 Bioavailability of nutrients in the mixture.
6 Postabsorptive disposition of the nutrients in the mixture.
7 Bioavailability of drugs given with the mixture.
8 Postabsorptive disposition of drugs.
9 Adverse nutritional and metabolic effects of drugs given with the mixture.

10   Drug–food and drug–nutrient incompatibilities associated with adverse reaction to the mixture.

COMPOSITION OF THE MIXTURE

The composition of enteral formula foods in common use is shown in Table 3.1A; the composition of parenteral fat emulsions and amino acids are shown in Tables 3.1B and 3.1C. When a formula is used as the sole source of nutrients, it must totally meet the needs of that patient. Under the following circumstances, the formula may not meet the patient's nutrient needs:

A.   When nutrients are "missing" from the mixture due to improper formulation by the pharmacist or manufacturer (chloride has been left out of infant formulas) (Roy 1982).

B.   When decision is made by the manufacturer to omit a nutrient from the mixture due to real or perceived stability problems. (Folate has been omitted from TPN formulas for this reason.)

C.   When the nutrient needs of the patient are altered because of disease or drug usage. (Patients who have had intestinal resections for inflammatory bowel disease and are receiving broad-spectrum antibiotics while on TPN need biotin since their ability to synthesize this vitamin in the gut is impaired.) Formulas may be too low in specific minerals for particular patient groups. For example, Isocal is insufficient in sodium for geriatric patients, who can exhibit inadequate sodium homeostasis. Hyponatremia has been reported in elderly demented patients on Isocal tube feeds (Rudman et al. 1986; Jamieson 1985). Risk of hyponatremia may be increased by concurrent administration of diuretics.

Formulas may also contain toxicants introduced as part of certain nutrient sources or due to contamination. Bone disease has been reported in patients on a home parenteral formula prepared using a casein. Aluminum in the casein inhibits bone mineralization (Howard and Michalek 1984). Examples of formulas (including their vitamin and mineral contents) used for tube feedings and for infant feeding are shown in Tables 3.2A–3.2C and Tables 3.3–3.11.

STABILITY OF THE MIXTURE

Instability of nutrients in formulas may be exacerbated by light exposure, temperature change, change in pH, and adsorption of nutrients onto the plastic containers or tubing.

Effects of lighting conditions on the degradation of B vitamins in parenteral formulas have been studied. Indirect and direct sunlight have been found to destroy riboflavin. Indirect sunlight destroys 47% of riboflavin-5′-phosphate and direct sunlight destroys 100% of this form of riboflavin in 8 hours. Direct sunlight photodegrades pyridoxine; 80% is destroyed in 8 hours. Under these same lighting conditions, folic acid and niacinamide are stable. It has also been reported that there is some degradation of thiamin with sunlight exposure (26% in 8 hours). However, in the study of effects of light on thiamin stability, other factors that could have affected stability of the vitamins were not controlled (Scheiner et al. 1981; Niemiec and Vanderveen 1984; Chen et al. 1983; Bowman and Nguyen 1983).

Thiamin stability in certain parenteral nutrition solutions is in doubt. Reduction in the thiamin content has been attributed to the sodium bisulfite used as an antioxidant in amino acid mixtures. One investigator reported that thiamin is stable in amino acid–dextrose solutions. Other investigators, including Scheiner et al. (1981), found large losses of thiamin in two different crystalline amino acid mixtures kept at 7°C and 23°C, respectively.

In the studies of Scheiner et al. (1981), the decomposing effect of light on thiamin was found to be negligible but lability of thiamin was found to be related to higher pH, longer storage, and higher temperatures of the amino acid solutions (Freamine III, 8.5%, and Travasol, 5.5%) to which the thiamin had been added. In a review of these studies, Niemiec and Vanderveen (1984) commented that the amino acid solutions used by these investigators did not simulate those employed in IV solutions. They also noted that replicate vitamin assays were not performed.

Vitamin A is another photodegradable vitamin for which losses are inevitable in enteral and parenteral solutions unless they are in photoprotected plastic containers. Indeed, it has been shown by Kishi et al. (1981) that destruction of 75% of the vitamin A in feeding solutions occurs in a 24-hour period.

INCOMPATIBILITY OF DRUGS AND NUTRIENTS WITH FORMULAS

Enteral or parenteral solutions may be incompatible with drugs and nutrients in the following ways:

Adsorption. Insulin as well as vitamin A may be adsorbed onto a plastic container or tubing (Goldberg and Levin 1978; Moorhatch and Chiou 1974).

Precipitation. Calcium and phosphate ions may be precipitated as dibasic calcium phosphate. The precipitation is favored by high molar concentration of the interacting ions, temperature elevation, increase

TABLE 3.1A ENTERAL FORMULAS

Product	Manufacturer	Calories/ml	Major Protein Source	Protein g/L (%)	Major CHO Source	CHO g/L (%)
Elemental						
Criticare HN	Mead Johnson	1.06	Enzymatically hydrolyzed casein and free amino acids	38.0 (14.0)	Maltodextrin, modified cornstarch	222 (83.0)
Nutrex Aminex	Nutrex	1	Free amino acids	38.3 (15.3)	Maltodextrin, modified starch	206 (82.2)
Travasorb HN	Travenol	1	Hydrolyzed lactalbumin	45.0 (18.0)	Glucose oligosaccharides	175 (70.0)
Travasorb Standard	Travenol	1	Hydrolyzed lactalbumin	30.0 (12.0)	Glucose oligosaccharides	190 (76.0)
Vital HN	Ross	1	Partially hydrolyzed whey & meat, soy, free amino acids	41.7 (16.7)	Hydrolyzed cornstarch, sucrose	185 (73.9)
Vivonex HN	Norwich Eaton	1	Free amino acids	44.4 (17.8)	Glucose oligosaccharides	210 (81.5)
Vivonex Standard	Norwich Eaton	1	Free amino acids	20.6 (8.2)	Glucose oligosaccharides	231 (92.4)
Vivonex T.E.N.	Norwich Eaton	1	Free amino acids (30% BCAA)	38.2 (15.3)	Maltodextrin, modified starch	206 (82.2)
Pepti-2000	Chesebrough-Pond's	1.1	Hydrolyzed lactalbumin	40 (16.0)	Maltodextrin	189 (79.5)
Blenderized						
Compleat Modified	Sandoz	1.07	Beef, calcium caseinate	43.0 (16.0)	Hydrolyzed cereal, fruit, vegetables	141 (54.0)
Compleat Regular	Sandoz	1.07	Beef, nonfat milk	43.0 (16.0)	Hydrolyzed cereal solids, fruit, vegetables, maltodextrin, lactose	128 (48.0)
Vitaneed	Chesebrough-Pond's	1	Pureed beef, sodium and calcium caseinates	35.0 (14.0)	Maltodextrin, fruit, vegetables	125 (50.0)
Milk-Based						
Carnation Instant Breakfast (with 8 oz. milk)	Carnation	1.06	Nonfat dry milk, calcium caseinate, sweet dairy whey	57.0 (22.0)	Sugar, corn syrup solids, lactose	133 (50.2)
Meritene Liquid	Sandoz	0.96	Concentrated sweet skim milk	57.6 (24.0)	Lactose, corn syrup solids, sucrose	110 (46.0)
Lactose-Free						
Enrich	Ross	1.1	Sodium & calcium caseinates, soy protein isolate	39.7 (14.5)	Hydrolyzed cornstarch, sucrose, soy polysaccharide	162 (55.0)
Ensure	Ross	1.06	Sodium & calcium caseinates, soy protein isolate	37.2 (14.0)	Corn syrup, sucrose	145 (54.5)
Ensure HN	Ross	1.06	Sodium & calcium caseinates, soy protein isolate	44.3 (16.7)	Corn syrup, sucrose	141 (53.0)
Ensure Plus	Ross	1.5	Sodium & calcium caseinates, soy protein isolate	54.9 (14.7)	Corn syrup, sucrose	200 (53.3)

TABLE 3.1A ENTERAL FORMULAS (CONT.)

Major Fat Source	Fat g/L (%)	Nonprotein Calories: gN	Volume to Meet 100% USRDA Vitamins & Minerals	mOsm/kg Water	Na mEq/ mg/L	K mEq/ mg/L	Form
Safflower oil	3.4 (3.0)	148:1	1887	650	27.5/632	33.9/1320	Liquid
Safflower oil	2.8 (2.5)	149:1	2000	600	20/461	20/782	Powder
MCT, sunflower oil	13.5 (12.0)	114:1	2000	560	40.1/922	30.0/1171	Powder
MCT, sunflower oil	13.5 (12.0)	184:1	2000	560	40.1/922	30.0/1171	Powder
Safflower oil, MCT	10.7 (9.4)	125:1	1500	460	20.3/466	34.2/1333	Powder
Safflower oil	0.9 (0.8).	125:1	3000	810	23.0/530	30.0/1174	Powder
Safflower oil	1.4 (1.3)	284:1	1800	550	20.4/468	30.0/1169	Powder
Safflower oil	2.8 (2.5)	139:1	2000	630	20.0/461	20.0/782	Powder
MCT, corn oil	10 (8.5)	131:1	1600	490	29/680	29/1150	Powder
Beef, corn oil	37.0 (30.0)	131:1	1500	300	29.1/670	35.9/1400	Liquid
Beef, corn oil	42.8 (36.0)	131:1	1500	405	56.5/1300	35.9/1400	Liquid
Partially hydrogenated soy oil	40.0 (36.0)	154:1	2000	310	22.0/500	32.0/1250	Liquid
Whole milk	31.0 (26.3)	89:1	1060 (except biotin)	677–715	42-52/ 969–1197	56.7–77.2/ 2212–3010	Powder
Corn oil	32.0 (30.0)	79:1	1250	505 (vanilla)	38.3/880	41.0/1600	Liquid
Corn oil	37.1 (30.5)	148:1	1391	480	36.7/845	40.0/1564	Liquid
Corn oil	37.1 (31.5)	153:1	1887	450	36.7/845	40.0/1564	Liquid
Corn oil	35.4 (30.1)	125:1	1321	470	40.4/930	40.0/1564	Liquid
Corn oil	53.2 (32.0)	146:1	1600	600	49.5/1139	54.1/2114	Liquid

(continued)

DIET AND DRUG INTERACTIONS

TABLE 3.1A ENTERAL FORMULAS (CONT.)

Product	Manufacturer	Calories/ml	Major Protein Source	Protein g/L (%)	Major CHO Source	CHO g/L (%)
Lactose-Free (cont.)						
Ensure Plus HN	Ross	1.5	Sodium & calcium caseinates, soy protein isolate	62.4 (16.7)	Corn syrup, sucrose	200 (53.3)
Entrition	Biosearch	1	Sodium & calcium caseinates	35.0 (14.0)	Maltodextrin	136 (54.4)
Fortical	Chesebrough-Pond's	1.5	Sodium caseinate, calcium caseinate	60 (16.0)	Maltodextrin, sucrose	180 (48.0)
Fortison	Chesebrough-Pond's	1.0	Sodium caseinate, calcium caseinate	40 (14.0)	Maltodextrin	120 (48.0)
Fortison L.S.	Chesebrough-Pond's	1.0	Sodium caseinate, calcium caseinate	40 (16.0)	Maltodextrin	120 (48.0)
Isocal	Mead Johnson	1.06	Calcium caseinate & sodium caseinate, soy protein isolate	34.2 (13.0)	Maltodextrin	133 (50.0)
Isocal HCN	Mead Johnson	2	Calcium caseinate & sodium caseinate	74.7 (15.0)	Corn syrup	224 (45.0)
Isolife	Navaco	1	Whey protein concentrate, soy protein	37.0 (15.0)	Hydrolyzed corn syrup	138 (55.0)
Isotein HN	Sandoz	1.2	Delactosed lactalbumin	68.0 (23.0)	Maltodextrin, monosaccharides	156 (52.0)
Magnacal	Chesebrough-Pond's	2	Sodium & calcium caseinates	70.0 (14.0)	Maltodextrin, sucrose	250 (50.0)
Newtrition High Nitrogen	Knight Medical	1.24	Sodium & calcium caseinate, soy protein isolate	60 (19.0)	Maltodextrin	160 (52.0)
Newtrition Isotonic	Knight Medical	1.06	Sodium & calcium caseinate, soy protein isolate	36 (14.0)	Maltodextrin	148 (56.0)
Newtrition (vanilla, chocolate)	Knight Medical	1.06	Sodium & calcium caseinate, soy protein isolate	36 (14.0)	Maltodextrin, sucrose, glucose solids	140 (52.0)
Nutrex Besure	Nutrex	1.06	Calcium & sodium caseinates, soy protein	37.2 (14.0)	Corn syrup solids, sucrose	145 (54.5)
Nutrex Encare	Nutrex	1.5	Calcium sodium caseinate, egg white	55 (14.4)	Corn syrup solids, sucrose	211.3 (55.6)
Nutrex Protamin	Nutrex	1.3	Calcium sodium caseinate	39.7 (12.6)	Corn syrup solids, sucrose	190 (60.3)
Osmolite	Ross	1.06	Sodium & calcium caseinate, soy protein isolate	37.2 (14.0)	Hydrolyzed cornstarch	145 (54.6)
Osmolite HN	Ross	1.06	Sodium & calcium caseinate, soy protein isolate	44.4 (16.7)	Hydrolyzed cornstarch	141 (53.3)
Precision HN Diet	Sandoz	1.05	Egg white solids	43.9 (16.7)	Maltodextrin, sucrose	216 (82.2)
Precision Isotonic Diet	Sandoz	1	Egg white solids	29.0 (12.0)	Glucose oligosaccharides, sucrose	144 (60.0)
Precision LR Diet	Sandoz	1.1	Egg white solids	26.0 (9.5)	Maltodextrin, sucrose	248 (89.2)
Pre-Fortison	Chesebrough-Pond's	0.5	Sodium caseinate, calcium caseinate	20 (16.0)	Maltodextrin	60 (48.0)
Resource	Sandoz	1.06	Sodium & calcium caseinate, soy protein isolate	37.2 (14.0)	Maltodextrin, sucrose	145 (54.5)
Sustacal Liquid	Mead Johnson	1	Calcium caseinate, soy protein isolate, sodium caseinate	61.2 (24.0)	Sugar (sucrose), corn syrup	140 (55.0)
Sustacal HC	Mead Johnson	1.5	Calcium and sodium caseinate	60.8 (16.0)	Corn syrup solids, sugar	190 (50.0)
Travasorb Liquid	Travenol	1.06	Sodium & calcium caseinate, soy protein isolate	37.1 (14.0)	Corn syrup solids, sucrose	144 (54.3)
Travasorb MCT Liquid	Travenol	1.5	Casein lactalbumin and other whey products	73.8 (20.0)	Maltodextrin	185 (50.0)
Travasorb MCT Powder	Travenol	1	Lactalbumin, potassium caseinate	49.3 (20.0)	Corn syrup solids	123 (50.0)
TwoCal HN	Ross	2	Sodium & calcium caseinate, soy protein isolate	83.5 (16.7)	Hydrolyzed cornstarch, sucrose	217 (43.2)

TABLE 3.1A ENTERAL FORMULAS (CONT.)

Major Fat Source	Fat g/L (%)	Nonprotein Calories: gN	Volume to Meet 100% USRDA Vitamins & Minerals	mOsm/kg Water	Na mEq/ mg/L	K mEq/ mg/L	Form
Corn oil	49.8 (30.0)	125:1	947	650	51.4/1181	46.5/1814	Liquid
Corn oil	35.0 (31.5)	154:1	2000	300	31.0/700	30.8/1200	Prefilled RTU pouch
Corn oil	60 (36.0)	131:1	1060	410/600	44/1020	44/1725	Liquid
Corn oil	40 (36.0)	131:1	1600	300	30/680	29/1150	Liquid
Corn oil	40 (36.0)	131:1	1600	240	9/200	29/1150	Liquid
Soy oil, MCT	44.3 (37.0)	167:1	1887	300	22.9/527	33.8/1316	Liquid
Soybean oil, MCT	90.7 (40.0)	145:1	1500	690	34.7/798	35.8/1396	Liquid
Corn oil	33.6 (30.0)	144:1	2000	300	31.3/720	25.6/1000	Liquid
Part. hydrog. soybean oil, MCT	34.0 (25.0)	86:1	1770	300	27.0/621	27.4/1070	Powder
Partially hydrogenated soy oil	80.0 (36.0)	154:1	1000	590	44/1000	32/1250	Liquid
MCT, corn oil	40 (29.0)	104:1	1240	300	26/598	26/1014	Liquid
Corn oil, MCT	36 (30.0)	154:1	1900	300	26/598	26/1014	Liquid
Corn oil	40 (34.0)	154:1	1900	450	26/598	26/1014	Liquid
Hydrogenated soybean oil, mono- and diglycerides	37.2 (31.5)	153:1	1887	450	36.8/846	40/1560	Powder
Soy oil, lecithin	50.7 (30.0)	146:1	1600	480	7.4/170	43.6/1700	Powder
Corn oil, lecithin	38.0 (27.1)	174:1	1180	450	7.4/170	43.6/1700	Powder
MCT, corn oil, soy oil	38.5 (31.4)	153:1	1887	300	23.8/546	26.0/1013	Liquid
MCT, corn oil, soy oil	36.8 (30.0)	125:1	1320	310	40.4/928	40.0/1561	Liquid
Part. hydrog. soy oil	1.3 (1.1)	125:1	2850	525	42.6/980	23.3/910	Powder
Partially hydrogenated soybean oil	30.0 (28.0)	183:1	1560	300	19.7/770	24.6/960	Powder
Part. hydrog. soy oil	1.6 (1.3)	239:1	1710	530 (vanilla)	30.4/700	22.6/880	Powder
Corn oil	20 (36.0)	131:1	1600	150	15/340	15/575	Liquid
Hydrogenated soy oil	37.2 (31.5)	154:1	1896	450	36.8/850	40.0/1560	Instant crystals
Partially hydrogenated soy oil	23.2 (21.0)	79:1	1080	625	40.9/928	53.0/2067	Liquid
Partially hydrogenated soybean oil	57.4 (34.0)	134:1	1200	650	36.7/844	37.6/1477	Liquid
Corn oil, partially hydrogenated soy oil	37.1 (31.5)	153:1	1887	488	32.1/738	32.5/1265	Liquid
MCT, safflower oil	48.5 (30.0)	100:1	1333	420	22.8/52	25.6/1480	Liquid
MCT, safflower oil	33.0 (30.0)	100:1	2000	312	15.2/350	25.6/1000	Powder
Corn oil, MCT	90.0 (40.1)	125:1	950	700	46/1050	59/2310	Liquid

(continued)

TABLE 3.1A ENTERAL FORMULAS (CONT.)

Product	Manufacturer	Calories/ml	Major Protein Source	Protein g/L (%)	Major CHO Source	CHO g/L (%)
Protein Modular Components						
Casec	Mead Johnson	3.7/g	Calcium caseinate	(95.0)		
Nutrisource Protein	Sandoz	4/g	Lactalbumin, egg white solids	(75.0)		(6.0)
Pro-Max	Nutrex	4.2/g	Whey protein	(71.3)		(9.5)
Pro-Mix R.D.P.	Navaco	3.6/g	Whey protein	(83.0)		(6.0)
Pro-Mod	Ross	4.2/g	Whey protein concentrate	(71.3)		(9.5)
Pro-Pac	Chesebrough-Pond's	4/g	Whey protein concentrate	(76.0)		(6.0)
Fat Modular Components						
High Fat Supplement	Navaco	6.12/g		(3.0)		(26.0)
MCT Oil	Mead Johnson	8.3				
Microlipid	Chesebrough-Pond's	4.5				
Nutrisource Lipid (long chain triglycerides)	Sandoz	2.2				
Nutrisource Lipid (medium chain triglycerides)	Sandoz	2				
Carbohydrate Modular Components						
Moducal	Mead Johnson	3.8/g			Maltodextrin	(100)
Nutrisource CHO	Sandoz	3.2			Deionized corn syrup solids	(100)
Polycose Liquid	Ross	2			Glucose polymers	(100)
Polycose Powder	Ross	3.8/g			Glucose polymers	(100)
Pure Carbohydrate Supplement	Navaco	4			Glucose polymers	(100)
Sumacal	Chesebrough-Pond's	4			Glucose polymers	(100)
Special Formulas						
Amin-Aid	Kendall McGaw	2	Essential amino acids plus histidine	19.4 (4.0)	Maltodextrin, sugar	366 (74.8)
Citrotein	Sandoz	0.66	Egg white solids	41.0 (25.0)	Sucrose, maltodextrin	122 (73.0)
Hepatic-Aid II	Kendall McGaw	1.2	Essential & nonessential amino acids	44.1 (15.0)	Maltodextrin, sucrose	169 (57.3)
Nutrex Broth	Nutrex	1.4	Sodium caseinate	43.4 (12.5)	Corn syrup solids	304 (87.5)
Nutrex CLD	Nutrex	0.95	Egg white solids	108.2 (46.8)	Sucrose	121.7 (52.6)
Nutrex Drink	Nutrex	0.78	Egg white solids	42 (20.9)	Corn syrup solids	158 (78.6)
Pulmocare	Ross	1.5	Sodium & calcium caseinate	62.4 (16.7)	Hydrolyzed cornstarch, sucrose	105 (28.1)
Ross SLD	Ross	0.7	Egg white solids	37.5 (21.4)	Sugar (sucrose), hydrolyzed cornstarch	137 (78.0)
Stresstein	Sandoz	1.2	Amino acids	69.7 (23.0)	Maltodextrin	173 (57.0)
Traumacal	Mead Johnson	1.5	Calcium caseinate	85.2 (22.0)	Corn syrup, sugar	142.5 (40.0)
Traum-aid HBC	Kendall McGaw	1	Amino acids (50% BCAA)	56.0 (22.4)	Maltodextrin	166 (66.4)
Travasorb Hepatic	Travenol	1.1	Crystalline L-amino acids	29.1 (10.6)	Glucose oligosaccharides	213 (77.5)
Travasorb Renal	Travenol	1.4	Crystalline L-amino acids	24.0 (6.9)	Glucose oligosaccharides	284 (81.1)

Source: Letsou, A. Nutritional Support Services, 7(3), March 1987.

TABLE 3.1A ENTERAL FORMULAS (CONT.)

Major Fat Source	Fat g/L (%)	Nonprotein Calories: gN	Volume to Meet 100% USRDA Vitamins & Minerals	mOsm/kg Water	Na mEq/ mg/L	K mEq/ mg/L	Form
	(5.0)				6.5/149.5/100 g	3.0/117/100 g	Powder
	(19.0)				29.0/670/100 g	36.0/1414/100 g	Powder
	(19.2)				8.6/197.8	25.1/978.9	Powder
	(11.0)				230 mg/100 g	828 mg/100 g	Powder
	(19.2)				8.6 mEq/196 mg/100 g	25.1 mEq/985 mg/100 g	Powder
	(18.0)				9.8 mEq/225 mg/100 g	12.8 mEq/500 mg/100 g	Powder
Part. hydr. coconut oil	(70.0)						Powder
Coconut oil	(100.0)						Liquid
Safflower oil	(100.0)						Liquid
Soybean oil	(100.0)						Liquid
Coconut oil	(100.0)						Liquid
							Powder
							Liquid
							Liquid
							Powder
							Powder
							Powder
Part. hydrog. soy oil	46.2 (21.2)	640:1	NA	700	Less than 15	Negligible	Powder
Part. hydrog. soy oil	1.6 (2.0)	76:1	180	480 (punch)	31.0/710	18.2/710	Powder
Part. hydrog. soy oil	36.2 (27.7)	340:1	—	560	Less than 15	Negligible	Powder
—	0	175:1	1065	350	17/391	27/1053	Powder
Mono- and di-glycerides	0.5 (0.6)	29:1	500	680	60/1380	38/1482	Powder
Mono- and di-glycerides	0.5 (0.5)	95:1	1060	450	29/667	45/1755	Powder
Corn oil	92.0 (55.2)	125:1	946	490	56.9/1308	48.7/1899	Liquid
	0.5 (0.6)	92:1	1200	545	36.3/835	21.4/835	Powder
MCT, soybean oil	27.3 (20.0)	97:1	2000	910	29.0/667	28.7/1121	Powder
Soybean oil, MCT	68.5 (38.0)	90:1	2000	550	51/1176	35.7/1386	Liquid
MCT, partially hydrogenated soybean oil	7.4 (6.7)	102:1	3000	760	23.0/529	30.0/1170	Powder
MCT, sunflower	14.6 (11.9)	211:1	2100	600	10.1/233	22.4/873	Powder
MCT, sunflower	18.6 (11.9)	340:1	2100	590	—	—	Powder

TABLE 3.1B COMPOSITION OF INTRAVENOUS FAT EMULSIONS

Product*,†,‡	Osmolarity (mOsm/L)	Glycerin (%)	Fatty Acid Content (%)			
			Linoleic	Linolenic	Oleic	Palmitic
Soybean Oil Based						
Intralipid 10%	260	2.25	50	9	26	10
Intralipid 20%	268	2.25	50	9	26	10
Soyacal 10%	280	2.21	49–60	6–9	21–26	9–13
Soyacal 20%	315	2.21	49–60	6–9	21–26	9–13
Travamulsion 10%	270	2.25	56	6	23	11
Travamulsion 20%	300	2.25	56	6	23	11
Safflower and Soybean Oil Based (50%/50%)						
Liposyn II 10%	320	2.5	65.8	4.2	17.7	8.8
Liposyn II 20%	340	2.5	65.8	4.2	17.7	8.8

*Intralipid is manufactured by Kabi-Vitrum, Soyacal by Greencross, Travamulsion by Travenol, and Liposyn by Abbott.
†All products with 10% lipid provide 1.1 kcal/ml. All products with 20% lipid provide 2.0 kcal/ml.
‡All products contain 1.2% egg yolk phospholipid emulsifier.

Source: Zeman, F.J. and Ney, D.M. *Applications of Clinical Nutrition.* Chapter 13, pp. 156 and 157. Englewood Cliffs, NJ: Prentice Hall, 1988.

in the pH of the solution, and addition of lipid (Niemiec and Vanderveen 1984).

Gel Formation. Liquid formulations of drugs, including suspensions and elixirs, can interact with enteral formulas when they are coadministered. Gel formation may clog the tube, and drug as well as nutrient may be precipitated out of solution. Liquid dosage forms of drugs that are incompatible with enteral formulas include those listed below (Cutie et al. 1983; Bauer 1982).

Dimetane elixir (A.H. Robbins)
Robitussin expectorant (A.H. Robbins)
Sudafed syrup (Burroughs-Wellcome)
Mellaril oral solution (Sandoz)
Thorazine concentrate (Smith Kline and French)
Cibalith-S syrup (Ciba)
Mandelamine suspension, Forte (Parke-Davis)
Feosol elixir (Menley and James)
KCL liquid (Barri)
Klorvess syrup (Dorsey)
MCT oil (Mead Johnson)
Neo-calglucon syrup (Dorsey)

TABLE 3.1C INTRAVENOUS CRYSTALLINE AMINO ACID SOLUTIONS

	Freamine HBC* (American McGaw)	Conventional Aminosyn 7% (Abbott)	Veinamine 8% (Cutter)	Hepatamine† (American McGaw)	Renamin‡ (Travenol)
Amino acid concentration (%)	6.9	7.0	8	8	6.5
Nitrogen (g/dl)	0.973	1.1	1.33	1.2	1
Essential amino acids (mg/dl)					
Isoleucine§	760	510	493	900	500
Leucine§	1370	660	347	1100	600
Lysine	410	510	667	610	450
Methionine	250	280	427	100	500
Phenylalanine	320	310	400	100	490
Threonine	200	370	160	450	380
Tryptophan	90	120	80	66	160
Valine§	880	560	253	840	820
Nonessential amino acids (mg/dl)					
Alanine	400	900	—	770	560
Arginine	580	690	749	600	630
Histidine	160	210	237	240	420
Proline	630	610	107	800	350
Serine	330	300	—	500	300
Tyrosine	—	44	—	—	40
Glycine	330	900	3387	900	300
Cysteine	<20			<20	
Glutamic acid			426		
Aspartic acid			400		
Electrolytes (mEq/L)					
Sodium	10		40	10	—
Potassium			30		
Chloride	<3	5.4	50	3	31
Magnesium			6		—
Acetate	57	105		62	60
Osmolarity (mOsm/L)	620	700	950	785	600

*Branched-chain amino acid–enriched stress formulation
†Hepatic failure formula
‡Renal failure formula
§Branched-chain amino acids

Source: Zeman, F.J. and Ney, D.M. *Applications of Clinical Nutrition.* Chapter 13, pp. 156 and 157. Englewood Cliffs, NJ: Prentice Hall, 1988.

TABLE 3.2A NUTRIENT COMPOSITION OF ROSS MEDICAL
NUTRITIONAL PRODUCTS

	Ensure	Ensure Plus
Energy	4435 kJ/L	6275 kJ/L
Energy distribution	(1060 kcal/L)	(1500 kcal/L)
Protein	14.0%	14.7%
Carbohydrate	54.5%	53.3%
Fat	31.5%	32.0%
Protein content	37.2 g/L	55.0 g/L
Protein source		
Sodium & calcium		
caseinates	87.5%	87.5%
Soy protein isolate	12.5%	42.5%
Calorie-nitrogen ratio	1.78%	171
Carbohydrate content	145 g/L	200 g/L
Carbohydrate source		
Lactose	0	0
Corn syrup solids	74%	74%
Sucrose	26%	26%
Fat content	37.2 g/L	53.3 g/L
Fat source		
Corn oil	100%	100%
Cholesterol	<20 mg/L	<25 mg/L
Renal solute load	329 mosm/L	479 mosm/L
Osmolality	450 mosm/kg H_2O	600 mosm/kg H_2O
Electrolytes		
Sodium	30.4 mmol L	47.8 mmol L
Potassium	30.7 mmol L	48.6 mmol L
Chloride	31.0 mmol L	45.1 mmol L

Usage: Liquid ready-to-feed formulas for use as supplement or diet of individuals with medical and surgical problems that risk nutrient depletion.

Source: Ensure and Ensure Plus, The Ross Medical Nutritional System. Ross Laboratories, Montreal, Canada. December 1983.

Dilantin oral suspensions (Parke-Davis)
Lanoxin elixir (Burroughs-Wellcome)

Among liquid formulations of drugs, syrups that are either strongly acidic or buffered to a pH of 4.0 or less cause immediate clumping when mixed with formulas. Mixtures may also increase in viscosity, particle size, and tackiness, increasing the risk of clogging the feeding tube. These problems cannot be avoided by diluting these syrups with water.

TABLE 3.2B APPROXIMATE MINERAL CONTENTS OF ROSS MEDICAL
NUTRITIONAL PRODUCTS

Mineral	Ensure		Ensure Plus	
	per liter	per 8500 kJ	per liter	per 8500 kJ
Sodium[a] (g)	0.72	1.38	1.06	1.44
Potassium[a] (g)	1.19	2.28	1.91	2.59
Chloride[a] (g)	1.10	2.11	1.60	2.17
Calcium (mg)	500	958	600	813
Phosphorus (mg)	500	958	600	813
Magnesium (mg)	200	383	300	406
Manganese (mg)	2.1	4	2.1	1.4
Iodine (μg)	85	163	106	144
Copper (mg)	1.1	2.1	1.61	2.18
Iron (mg)	9.5	18.2	14.3	19.4
Zinc (mg)	16	30.7	23.7	42.9

[a]Amounts are average amounts listed on labels.

Source: Ensure and Ensure Plus, The Ross Medical Nutritional System. Ross
Laboratories, Montreal, Canada. December 1983. Figures are estimated safe
and adequate daily dietary intakes.

It is therefore strongly recommended that such syrups not be used with
enteral products.

It may be possible to substitute compatible for incompatible drug
formulations, but this is not always feasible. For example, substitution
of pulverized iron preparations for soluble iron products, which cause
gelling of formulas, may be impossible because of the difficulty in
pulverizing the iron preparations (Altman et al. 1984).

Recommendations concerning the use of drugs with enteral formulas
are as follows:

A. Drugs should be given to patients on enteral feeding regimens
    as single bolus doses one hour or more after the flow of formula
    has temporarily been discontinued and the enteral feeding tube
    has been washed out with water.

**TABLE 3.2C APPROXIMATE VITAMIN CONTENTS OF ROSS MEDICAL
NUTRITIONAL PRODUCTS**

	Ensure		Ensure Plus	
Vitamin	per liter	per 8500 kJ	per liter	per 8500 kJ
Vitamin A (μg RE)[a]	793	1520	793	1074
Vitamin D (μg cholecalciferol)[b]	5	9	5	6.8
Vitamin E (mg α-TE)[c]	16	31	22	30
Vitamin K (μg)	38	73	55	74
Vitamin C (Ascorbic acid) (mg)	159	305	160	217
Folacin (μg)	200	383	210	284
Vitamin B_1 (Thiamin) (mg)	1.6	3.1	2.6	3.5
Vitamin B_2 (Riboflavin) (mg)	1.8	3.4	2.7	3.7
Vitamin B_6 (Pyridoxine) (mg)	2.1	4	3.1	4.2
Vitamin B_{12} (Cyanocobalamin) (μg)	6.3	12	8.8	11.9
Niacin[d] (mg)	21.0	40.2	31.7	42.9
Biotin (μg)	200	383	317	429
Pantothenic Acid (mg)	5.3	10.2	8.4	11.4
Choline (mg)	595	1140	510	690

[a]RE = retinol equivalent; 1 RE = 1 μg retinol or 6 μg B-carotene.
[b]10 μg cholecalciferol = 400 IU vitamin D.
[c]α-TE = α-tocopherol equivalent; 1 α-TE = 1 mg dl-α-tocopherol.
[d]NE = niacin equivalent; 1 NE = 1 mg niacin or 60 mg dietary tryptophan.

Source: Ensure and Ensure Plus, The Ross Medical Nutritional System. Ross
Laboratories, Montreal, Canada. December 1983. Figures are estimated safe
and adequate daily dietary intakes.

TABLE 3.3 SUSTACAL LIQUID (MEAD JOHNSON)

Nutrient	940 mL
Protein, g	57.2
Fat, g	22.4
Carbohydrate, g	132
Energy kcal	960
MJ	4
Vitamin A, IU	4416
RE	1336
Vitamin D, IU	352
mcg cholecalciferol	8.4
Vitamin E, IU	26.4
mg d-α-tocopherol	16.8
Vitamin C (ascorbic acid), mg	92.8
Thiamin (Vitamin B_1), mg	1.32
Riboflavin(Vitamin B_2), mg	1.6
Niacin, mg	18.8
NE	28.4
Vitamin B_6, mg	1.88
Folic acid (folacin), mg	0.09
Pantothenic acid, mg	10.4
Vitamin B_{12}, mcg	0.008
Sodium mg	1036
mEq	44.8
Potassium, mg	1644
mEq	42.2
Chloride, mg	1568
mEq	44.4
Calcium, g	1.42
Phosphorus, g	1.27
Magnesium, mg	140
Iron, mg	16
Iodine, mcg	128
Copper, mg	1.32
Zinc, mg	8
Manganese, mg	2.64
Choline, mg	—

Usage: Used to supplement or meet dietary needs of patients who have difficulty meeting their nutrient needs with a normal diet.

Source: Enteral Nutrition Handbook. Mead Johnson, Canada.

TABLE 3.4 ISOCAL (MEAD JOHNSON)

Nutrient	4 cans = 940 mL
Protein, g	32
Fat, g	41.2
Carbohydrate, g	125.2
Energy, kcal	1000
MJ	3.8
Vitamin A, IU	2480
RE	752
Vitamin D, IU	198
mcg cholecalciferol	4.8
Vitamin E, IU	37.6
mg d-α-tocopherol	25.2
Vitamin C (ascorbic acid), mg	150.4
Thiamin (Vitamin B_1), mg	1.9
Riboflavin (Vitamin B_2), mg	2.2
Niacin, mg	24.4
NE	36
Vitamin B_6, mg	2.4
Folic acid (folacin), mg	0.2
Pantothenic acid, mg	12.4
Vitamin B_{12}, mcg	8
Vitamin K, mcg	120
Biotin, mcg	152
Sodium, mg	500
mEq	21.6
Potassium, mg	1240
mEq	32
Chloride, mg	996
mEq	28
Calcium, g	0.6
Phosphorus, g	0.5
Magnesium, mg	196
Iron, mg	8.9
Iodine, mcg	76
Copper, mg	1.0
Zinc, mg	10
Manganese, mg	2.5
Choline, mg	244

Usage: Use to meet food-energy and nutrient needs of tube-fed patients. Canadian RDNI not established for some nutrients.

Source: Enteral Nutrition Handbook. Mead Johnson, Canada.

TABLE 3.5 FLEXICAL (MEAD JOHNSON)

Nutrient	100 mL	1000 mL
Protein, g	2.3	22.5
Fat, g	3.4	34
Carbohydrate, g	15	152
Energy, kcal	100	1000
MJ	0.4	4.2
Vitamin A, IU	250	2500
RE	76	758
Vitamin D, IU	20	200
mcg cholecalciferol	0.5	5
Vitamin E, IU	2.2	22
mg d-α-tocopherol	1.5	14.8
Vitamin C (ascorbic acid), mg	15	150
Thiamin (Vitamin B_1), mg	0.19	1.9
Riboflavin (Vitamin B_2), mg	0.22	2.2
Niacin, mg	2.5	25
NE	3.1	31
Vitamin B_6, mg	0.3	2.5
Folic acid (folacin), mg	0.02	0.2
Pantothenic acid, mg	1.2	12
Vitamin B_{12}, mcg	0.8	7.5
Vitamin K, mcg	13	125
Biotin, mcg	15	150
Sodium mg	35	350
mEq	1.5	15.2
Potassium, mg	130	1259
mEq	3.3	32.1
Chloride, mg	100	1000
mEq	2.8	28.2
Calcium, g	0.06	0.6
Phosphorus, g	0.05	0.5
Magnesium, mg	20	200
Iron, mg	1	9
Iodine, mcg	8	75
Copper, mg	0.1	1
Zinc, mg	1	10
Manganese, mg	0.3	2.5
Choline, mg	25	250

Usage: Use when an elemental diet is required, particularly in the short bowel syndrome and with pancreatic causes of maldigestion and malabsorption.

Source: Enteral Nutrition Handbook. Mead Johnson, Canada.

TABLE 3.6 SUSTAGEN (MEAD JOHNSON)

Nutrient	900 mL**
Protein, g	105
Fat, g	15
Carbohydrate, g	300
Energy, kcal	1750
MJ	7.4
Vitamin A, IU	5000
RE	1515
Vitamin D, IU	400
mcg cholecalciferol	10
Vitamin E, IU	45
mg d-α-tocopherol	30
Vitamin C (ascorbic acid), mg	300
Thiamin (Vitamin B$_1$), mg	3.8
Riboflavin (Vitamin B$_2$), mg	4.3
Niacin, mg	50
NE	—
Vitamin B$_6$, mg	5
Folic acid (folacin), mg	0.4
Pantothenic acid, mg	25
Vitamin B$_{12}$, mcg	15
Vitamin K, mcg	250
Biotin, mcg	300
Sodium, mg	1200
mEq	39.5
Potassium, mg	3200
mEq	82.1
Chloride, mg	—
Calcium, g	3.2
Phosphorus, g	2.4
Magnesium, mg	400
Iron, mg	18
Iodine, mcg	150
Copper, mg	2
Zinc, mg	20
Manganese, mg	5
Choline, mg	500

Usage: Use as dietary supplement.

**Approximate quantity prepared with 1 can (454 g) SUSTAGEN plus 682 mL (1.9 kcal/mL)
—Not present or unknown.

Source: Enteral Nutrition Handbook. Mead Johnson, Canada.

TABLE 3.7 SIMILAC/ISOMIL (ROSS)

Approximate Analysis (wt/liter)

Protein	20	g
Fat	36	g
Carbohydrate	68	g
Minerals	3.8	g
Calcium	0.7	g
Phosphorus	0.5	g
Sodium	0.3	g
Potassium	0.71	g
Chloride	0.53	g
Magnesium	50	mg
Iron	12	mg
Zinc	5	mg
Copper	0.5	mg
Iodine	0.15	mg
Water	901.6	g
Calories per fl oz	20	
Calories per 100 ml	68	

Vitamins Per Liter

Vitamin A	2500	IU
Vitamin D	400	IU
Vitamin E	15	IU
Vitamin C	55	mg
Vitamin B_1	0.4	mg
Vitamin B_2	0.6	mg
Vitamin B_6	0.4	mg
Niacin (mg equiv)	9	
Folic acid	0.1	mg
Vitamin B_{12}	3	mcg
Pantothenic acid	5	mg
Biotin	0.15	mg
Vitamin K_1	0.15	mg

Usage: Use to feed young infants who cannot be breastfed, who have sensitivity to cow's milk protein, and older infants with similar sensitivity. The addition of iron to this formula conforms to the recommendation of the Committee on Nutrition of the American Academy of Pediatrics.

Source: Physicians' Desk Reference, 30th Ed. Oradell, NJ: Medical Economics Co., Litton Industries.

TABLE 3.8 NUTRAMIGEN POWDER (MEAD JOHNSON)

One quart of NUTRAMIGEN formula (4.9 oz. NUTRAMIGEN Powder) supplies 640 kilocalories and the following vitamins and minerals:

Vitamin A, I.U.	1600
Vitamin D, I.U.	400
Vitamin E, I.U.	10
Vitamin C, mg	50
Folic acid, mcg	100
Thiamin, mg	0.5
Riboflavin, mg	0.6
Niacin, mg	8
Vitamin B_6, mg	0.4
Vitamin B_{12}, mcg	2
Biotin, mg	0.05
Pantothenic acid, mg	3
Vitamin K_1, mcg	100
Choline, mg	85
Calcium, mg	600
Phosphorus, mg	450
Iodine, mcg	45
Iron, mg	12
Magnesium, mg	70
Copper, mg	0.6
Zinc, mg	4
Manganese, mg	1
Chloride, mg	450
Potassium, mg	650
Sodium, mg	300

Usage: Hypoallergenic formula food that can be used to feed infants and children who are allergic to intact food proteins.

Source: Physicians' Desk Reference, 30th Ed. Oradell, NJ: Medical Economics Co., Litton Industries.

B. Alternatively, if swallowing is unimpaired, the drug may be given by mouth.

C. Coordination and monitoring of drug regimens with enteral formulas should be carried out by hospital pharmacies.

ALLERGENICITY

The allergenicity of ingredients of enteral and parenteral formulas may be due to the presence of haptenic substances, be they nutrients or intentional additives (Parker 1975). Thiamin is an example of a nutrient which in sensitized individuals can elicit an allergic reaction. Sodium bisulfite in amino acid solutions is an intentional additive

TABLE 3.9 PORTAGEN (MEAD JOHNSON)

One packed level 8-oz. measuring cup (135.6 g) of Portagen powder supplies 640 calories, 22.4 g protein, 30.5 g fat, 73.6 g carbohydrate, and the following vitamins and minerals:

	One Quart 20 Cal./ fl. oz.	One Quart 30 Cal./ fl. oz.
Vitamin A, I.U.	5000	7500
Vitamin D, I.U.	500	750
Vitamin E, I.U.	20	30
Vitamin C (Ascorbic acid), mg	50	75
Folic acid (Folacin), mg	0.1	0.15
Thiamin (Vitamin B_1), mg	1	1.5
Riboflavin (Vitamin B_2), mg	1.2	1.8
Niacin, mg	13	20
Vitamin B_6, mg	1.3	2
Vitamin B_{12}, mcg	4	6
Biotin, mg	0.05	0.075
Pantothenic acid, mg	6.7	10
Vitamin K_1, mg	0.1	0.15
Choline, mg	85	125
Calcium, g	0.6	0.9
Phosphorus, g	0.45	0.68
Iodine, mcg	45	70
Iron, mg	12	18
Magnesium, mg	130	200
Copper, mg	1	1.5
Zinc, mg	6	9
Manganese, mg	2	3
Chloride, mg	550	825
Potassium, g	0.8	1.2
Sodium, g	0.3	0.45

Usage: For use in feeding children and adults with defects in the intraluminal hydrolysis of fats, or mucosal defects in fat absorption, or lymphatic defects in the transport of fat with intestinal lymphate obstruction, in whom conventional food fats are not well tolerated.

Source: Physicians' Desk Reference, 30th Ed. Oradell, NJ: Medical Economics Co., Litton Industries.

TABLE 3.10 PREGESTIMIL (MEAD JOHNSON)

One quart of PREGESTIMIL formula (20 kcal./fl. oz.) supplies the following vitamins and minerals:

		% U.S. RDA Children Under 4 Years of Age
Vitamin A, I.U.	1600	64
Vitamin D, I.U.	400	100
Vitamin E, I.U.	10	100
Vitamin C, mg	50	125
Folic acid, mcg	100	50
Thiamin, mg	0.5	71
Riboflavin, mg	0.6	75
Niacin, mg	8	89
Vitamin B_6, mg	0.4	57
Vitamin B_{12}, mcg	2	67
Biotin, mg	0.05	33
Pantothenic acid, mg	3	60
Vitamin K_1, mcg	100	*
Choline, mg	85	*
Calcium, mg	600	75
Phosphorus, mg	450	56
Iodine, mcg	45	64
Iron, mg	12	120
Magnesium, mg	70	35
Copper, mg	0.6	60
Zinc, mg	4	50
Manganese, mg	1	*
Chloride, mg	450	*
Potassium, mg	650	*
Sodium, mg	300	*

Usage: Use in the feeding of infants and children with severe disaccharidase deficiencies.

*U.S. Recommended Daily Allowance (U.S. RDA) has not been established.
Source: Physicians' Desk Reference, 30th Ed. Oradell, NJ: Medical Economics Co., Litton Industries.

which can cause allergic reactions (Settipane 1984). Tartrazine, a yellow dye component of some drugs, is another common allergenic substance that may be administered inadvertently to patients on enteral hyperalimentation (Freedman 1977). A coadministered drug itself can evoke an allergic reaction. Whether the hapten is a nutrient, additive, or drug, it combines with an endogenous protein to form a complete antigen. Though combination of the hapten with a protein is usual,

TABLE 3.11 LOFENALAC (MEAD JOHNSON)

	Per Qt.
Protein, g	†
Vitamin A, I.U.	1600
Vitamin D, I.U.	400
Vitamin E, I.U.	10
Vitamin C (Ascorbic acid), mg	50
Folic acid (Folacin), mcg	100
Thiamin (Vitamin B_1), mg	0.5
Riboflavin (Vitamin B_2), mg	0.6
Niacin, mg	8
Vitamin B_6, mg	0.4
Vitamin B_{12}, mcg	2
Biotin, mg	0.05
Pantothenic acid, mg	3
Vitamin K_1, mcg	100
Choline, mg	85
Calcium, mg	600
Phosphorus, mg	450
Iodine, mcg	45
Iron, mg	12
Magnesium, mg	70
Copper, mg	0.6
Zinc, mg	4
Manganese, mg	1
Chloride, mg	450
Potassium, mg	650
Sodium, mg	300

Usage: Use in infants and children with phenylketonuria.

†The protein is incomplete since it contains an inadequate amount of the essential amino acid, phenylalanine, for normal growth. With added phenylalanine, the PER is greater than that of casein.

Source: Physicians' Desk Reference, 30th Ed. Oradell, NJ: Medical Economics Co., Litton Industries.

penicillin and other drugs may form antigenic complexes with amino acids in the nutrient mixture which is being administered (Niemiec and Vanderveen 1984).

Allergenicity may also be related to proteins being administered, whether these consist of the macronutrient components of the mixture or coadministered insulin, for example.

Adverse outcomes of giving allergenic substances include urticaria, asthma, anaphylactic shock, epidermal necrolysis, or other drug eruptions. These reactions may be life-threatening. The risks of severe allergic reactions depend on the rapid entry of the allergen into the circulation, the dose, and the condition of the patient (Menne and Hjorth 1982).

BIOAVAILABILITY OF NUTRIENTS

The bioavailability of nutrients in enteral formulas may be impaired due to the following conditions:

A. Instability of the formula.
B. Incompatibility of the formula with a drug or nutrient supplement, e.g., iron.
C. Introduction of the formula at a site distal to the site of optimal nutrient absorption. Enteral formulas delivered into the lower jejunum may be less efficient as nutrient sources for this reason.
D. Formula-induced osmotic diarrhea. Glucose polymers in the formula can induce such diarrhea, particularly in infants.
E. Adsorption of nutrients onto formula constituents, e.g., fiber sources. Psyllium seed added to formula to prevent constipation may reduce the bioavailability of riboflavin (Kalkwarf et al. 1987).

Lactose in enteral formulas may reduce calcium absorption in lactase deficient patients (Cochet et al. 1983).

POSTABSORPTIVE DISPOSITION OF NUTRIENTS

The postabsorptive disposition of nutrients is affected by the administration of enteral and parenteral formulas to malnourished patients. When formulas provide liberal amounts of food energy, protein is spared and less exogenous protein is used as an energy source (Goodwin and Wilmore 1983).

BIOAVAILABILITY OF DRUGS

The bioavailability of phenytoin is reduced by coadministration of therapeutic doses of folic acid (>5.0 mg) or of vitamin B_6. The reduction in phenytoin absorption produced by these two vitamins was observed in patients on normal diets (Baylis et al. 1971; Mattson et al. 1973; Roe 1985). Since the vitamin-induced decrease in the bioavailability of phenytoin can result in loss of seizure control, it is recommended that enthusiasm for administration of folic acid or vitamin B_6 supplement with enteral formulas be balanced with the knowledge that these vitamins can reduce the therapeutic efficacy of the anticonvulsant.

In patients receiving continuous nasogastric feedings with Osmolite or Isocal, the absorption of phenytoin suspension or capsule may be reduced as much as 70% (Bauer 1982; Hatton 1984). It is recommended that phenytoin–formula interactions should be managed by flushing and clamping the nasogastric tube two hours before the drug dose and flushing again two hours after the dose, before restarting the formula infusion.

Vitamin K in enteral formulas or liquid nutrient supplements can interfere with the anticoagulant effect of coumarin drugs. It has been considered that this effect is explained by a post-absorptive interaction (Riley and Rytand 1980).

Administration of enteral formula containing a hemicellulose and pectin-containing fiber source enhances the absorption of theophylline (Mayne and Roe 1987). For patients on phenytoin who have experienced reduction in drug absorption due to continuous enteral feeding, it is *not* recommended to increase the drug dosage. Rather, one should continue the drug schedule and change the feeding regimen to balance feeding with flushing of the tubes before and after drug administration because of the risk of drug toxicity with higher dosage (Saklad et al. 1986; Ozuna and Friel 1984).

POSTABSORPTIVE DISPOSITION OF DRUGS

The rate of drug metabolism may be increased by the administration of a high-protein enteral or parenteral formula. High-protein diets have been shown to increase the rate of metabolism of antipyrine and theophylline (Kappas et al. 1976).

ADVERSE NUTRITIONAL AND METABOLIC EFFECTS OF ENTERAL AND PARENTERAL HYPERALIMENTATION

Omission of Nutrients from the Formula. Chloride deficiency has occurred in infants fed a chloride-deficient formula (Roy 1982).

Biotin deficiency has been reported in patients with inflammatory bowel disease who have received broad-spectrum antibiotics while on TPN. Biotin deficiency in these patients is the outcome of zero biotin intake (none in the TPN mixture), and destruction by antibiotics of the normal intestinal microflora that synthesize biotin (Levenson 1983; McClain et al. 1982; Mock et al. 1981).

Unusual Nutritional Requirements of Patients on TPN. Carnitine deficiency may develop in patients on TPN. Patients on enteral formula diets or TPN may require carnitine supplements. Carnitine is normally synthesized in the body from the amino acids methionine and lysine.

Enteral diets in which the main protein source is soy protein isolate, casein, or egg-white protein contain little or no carnitine and may provide too little methionine for its endogenous synthesis. Solutions used in TPN do not contain carnitine (Borum 1983).

Carnitine is required for the transport of long-chain fatty acids into mitochondria. In carnitine deficiency this function is impaired, normal fatty acid oxidation cannot take place, and energy metabolism is severely compromised. Carnitine is necessary to the normal function of cardiac muscle, skeletal muscle, and the liver. It may protect against toxic (drug-induced) myopathies as well as against development of a fatty liver.

It has been postulated that patients receiving enteral formula diets or TPN may also be at risk for taurine deficiency. Taurine may be required by infants, especially premature infants, for development of normal vision; evidence is based in part on studies of taurine deficiency in kittens. Taurine is normally obtained from animal protein foods and is synthesized from sulfur-containing amino acids (Malloy et al. 1981). It should, however, be noted that high intake of taurine may cause itching and may exacerbate psoriasis (Roe 1965).

Other special nutrient requirements of patients on TPN have been identified by Howard and Michalek (1984). These authors emphasize a high requirement for essential fatty acids and calcium, and increased requirements for zinc and copper.

Adverse metabolic effects of TPN and elemental diets include the following:

Complications Related to Carbohydrate.

1 Glucose has an irritant effect on veins.
2 Fructose can predispose patients to lactic acidosis and hypophosphatemia.
3 Sorbitol is converted to fructose and therefore imposes the risk of similar complications. Sorbitol can also cause osmotic diarrhea.
4 Ethanol predisposes to lactic acidosis and gout.
5 Hypoglycemia may occur when a high-dose IV glucose infusion is suddenly stopped.
6 Hyperglycemia may develop if the patient is unable to utilize glucose efficiently. Hyperosmolar coma may occur as a complication of untreated hyperglycemia.
7 Respiratory distress may occur with excessive CO_2 production secondary to a high carbohydrate load.

Complications Related to Lipid.   Side effects, which have been reported in relation to use of Intralipid, include fever, impaired pulmonary function, depressed cell-mediated immunity, and impaired utilization with lipemia.

Complications Related to Amino Acids.   Use of protein hydrolysates can produce elevated blood urea, hyperammonemia, and hepatic encephalopathy. The latter is a particular risk in patients with alcoholic liver disease. Hyperchloremic acidosis may occur with an excessive chloride load because of the kidney's reduced ability to excrete hydrogen ions.

Complications Related to Minerals and Trace Elements.   Hypo- and hyperphosphatemia may occur. Increased utilization of nutrients in the catabolic patient increases phosphate needs which can deplete plasma phosphate. Symptoms associated with phosphate depletion include muscle weakness, tremor, paresthesias, hemolysis, confusion, and coma. Excess phosphate administration can lead to metastatic calcification when hypercalcemia is present.

Magnesium deficiency may develop with consequent muscle weakness and tetany (Greenhall 1982).

ADVERSE DRUG REACTIONS

Adverse drug reactions include allergic reactions discussed above and hemorrhagic phenomena in patients receiving cephalosporin antibiotics that have an anti–vitamin K effect. Hemorrhage from this cause is more likely to occur in patients on TPN who have been receiving formula without addition of vitamin K (*Nutr. Rev.* 1984). Maintenance of oral anticoagulation without high risk of hemorrhagic side effects in patients requiring parenteral or enteral support requires careful selection of the anticoagulating agent and a frequently monitored therapeutic schedule. For patients requiring such forms of nutritional support, who also have evidence of vitamin K malabsorption, low-dose warfarin maintenance is recommended (Gimmon 1987).

REFERENCES

ALTMAN, E., A.J. CUTIE, and M. SCHWARTZ. 1984. Compatibility of enteral products with commonly employed drug additives. *Nutr. Supp. Svc.* 4:8.

BAUER, L.A. 1982. Interference of oral phenytoin absorption in continuous nasogastric feedings. *Neurology* 32: 570.

BAYLIS, E.M., J.M. CRAWLEY, and J.M. PREECE, et al. 1971. Influence of folic acid on blood phenytoin levels. *Lancet* 1: 62–64.

BORUM, P.R. 1983. Carnitine. *Ann. Rev. Nutr.* 3: 233.

BOWMAN, B.B. and P. NGUYEN. 1983. Stability of thiamin in parenteral nutrition solutions. *J. Parent. Ent.* 7: 567.

CHEN, M.F., H.W. BOYCE, and L. TRIPLETT. 1983. Stability of the B vitamins in mixed parenteral nutrition solutions. *J. Parent. Ent. Nutr.* 7: 462.

COCHET, T., A. JUNG, M. GRIESSEN, P. BARTHOLDI, P. SCHALLER, and A. DONATH. 1983. Effects of lactose on intestinal calcium absorption in normal and lactase-deficient subjects. *Gastroenterology* 84: 935.

CUTIE, A.J., E. ALTMAN, and L. LENKEL. 1983. Compatibility of enteral products with commonly employed drug additives. *J. Parent. Ent. Nutr.* 7: 186.

FREEDMAN, B.J. 1977. Asthma induced by sulfur dioxide benzoate and tartrazine contained in orange drinks. *Clin. Allergy* 7: 407.

GIMMON, Z. 1987. Oral anticoagulant therapy in patients who require nutritional support. *J. Parent. Ent. Nutr.* 11: 102.

GOLDBERG, N.J. and S.R. LEVIN. 1978. Insulin adsorption to an in-line membrane filter. Letter. *New Eng. J. Med.* 298: 1480.

GOODWIN, C.W. and D.W. WILMORE. 1983. Enteral and Parenteral Nutrition. In *Manual of Clinical Nutrition.* 31.1–31.36. Pleasantville, NJ: Nutrition Publications Inc.

GRANT, A. 1982. *Basic Nutritional Background in Enteral and Parenteral Nutrition.* 21. Oxford and London: Blackwell.

GREENHALL, M. 1982. Metabolic complications. In *Enteral and Parenteral Nutrition,* ed. A. Grant and E. Todd. 97. Oxford and London: Blackwell.

HATTON, R.C. 1984. Dietary interaction with phenytoin. *Clin. Pharm.* 3: 110–11.

HOWARD, L. and A.V. MICHALEK. 1984. Home parenteral nutrition (HPN). *Ann. Rev. Nutr.* 4: 69.

JAMIESON, J.J. 1985. Hyponatraemia. *Brit. Med. J.* 290: 1723–28.

KAPPAS, A., K.E. ANDERSON, A.H. CONNEY, and A.P. ALVARES. 1976. Influence of dietary protein and carbohydrate on antipyrine and theophylline metabolism in man. *Clin. Pharmacol. Ther.* 20: 643.

KISHI, H., A. YAMAJI, K. KATAOKA, Y. FUJII, K. NISHIKAWA, N. OHNISHI, E. HIRAOKA, A. OKADA, and C.W. KIM. 1981. Vitamin A and E requirements during total parenteral nutrition. *J. Parent. Ent. Nutr.* 5: 420–23.

LEVENSON, J.L. 1983. Biotin responsive depression during hyperalimentation. *J. Parent. Ent. Nutr.* 7: 181.

MALLOY, M.H., D.K. RASSIN, G. GAULL, and W.C. HEIRD. 1981. Development of taurine metabolism in Beagle pups: effects of taurine-free total parenteral nutrition. *Biol. Neonate.* 40: 1.

MATTSON, R.H., B.B. GALLAGHER, E.H. REYNOLDS, and D. GLASS. 1973. Folate therapy in epilepsy. A controlled study. *Arch. Neuro.* 29: 78–81.

MAYNE, J., and D.A. ROE. 1987. Effects of varied fiber level in enteral formula upon the xenobiotic metabolizing systems of rat intestinal mucosa. Abstract. *Fed. Proc.* 46: 1168.

McCLAIN, C.J., H. BAKER, and G.R. OMSTAD. 1982. Biotin deficiency in an adult during home parenteral nutrition. *J. Am. Med. Assoc.* 247: 3116.

MENNE, T. and N. HJORTH. 1982. Reactions from systemic exposure to contact allergens. *Seminars in Dermatol.* 1: 15.

MOCK, D.M., A.A. DELORIMER, W.M. LIEBMAN, L. SWEETMAN, and H. BAKER. 1981. Biotin deficiency: an unusual complication of parenteral alimentation. *New Eng. J. Med.* 304: 820.

MOORHATCH, P. and W.L. CHIOU. 1974. Interactions between drugs and plastic intravenous fluid bags. *Am. J. Hosp. Pharm.* 31: 72.

NIEMIEC, P.W., Jr. and T.W. VANDERVEEN. 1984. Compatibility considerations in parenteral nutrient solutions. *Am. J. Hosp. Pharm.* 41: 893.

NUTRITION REVIEWS. 1984. Editorial. New examples of vitamin K–drug interaction. *Nutr. Rev.* 42: 161.

OZUNA, J. and P. FRIEL. 1984. Effect of enteral tube feeding on serum phenytoin levels. *J. Neurosurg. Nurs.* 16: 289–91.

PARKER, C.W. 1975. Drug therapy: Drug allergy (First of three parts). *New Eng. J. Med.* 292: 511.

RILEY, R. and D. RYTAND. 1980. Resistance to warfarin due to unrecognized vitamin K supplementation. *New Eng. J. Med.* 303: 160.

ROE, D.A. 1965. Nutrient Requirements in Psoriasis. *New York State J. Med.* 65: 1319.

———. 1985. *Drug-Induced Nutritional Deficiencies,* 2nd ed. 251–52. Westport, CT: AVI.

ROE, D.A., H. KALKWARF, and J. STEVENS. 1988. Effect of fiber supplements on the apparent absorption of pharmacological doses of riboflavin. *J. Am. Dietet. A.* 88: 211–13.

ROY, S. 3rd. 1982. Metabolic alkalosis from chloride deficient infant formula. In *Adverse Effects of Foods.* ed. E.F. Jelliffe and D.B. Jelliffe. 575. New York: Plenum Press.

RUDMAN, D., D. RACETTE, I.W. RUDMAN, D.E. MATTSON, and P.R. ERVE. 1986. Hyponataremia in tube-fed elderly men. *J. Chron. Dis.* 39: 73–80.

SAKLAD, J.J., R.H. GRAVES, and W.P. SHARP. 1986. Interaction of oral phenytoin with enteral feedings. *J. Parent. Ent. Nutr.* 10: 322–23.

SCHEINER, J.M., M.M. ARAUJO, and E. DERITTER. 1981. Thiamín destruction by sodium bisulfite in infusion solutions. *Am. J. Hosp. Pharm.* 38: 1911.

SETTIPANE, G.A. 1984. Adverse reactions to sulfites in drugs and foods. *J. Am. Acad. Dermatol.* 10: 1077.

Chapter 4

DRUG-INDUCED REACTIONS TO ALCOHOL AND FOOD

ADVERSE EFFECTS OF ALCOHOL (ETHANOL) IN DRUG FORMULATIONS

Many over-the-counter (OTC) drugs as well as a few prescription drugs contain alcohol (Bailey 1975). Drug mixtures containing alcohol include tonics, bronchodilators, cough medicines, cold cures, diarrhea medicines, and analgesics. Tonics containing alcohol that may be taken as sources of iron and vitamins include Niferex Forte Elixir (Central Pharmacal) and Nu-Iron-Plus Elixir (Mayrand), both containing 10% alcohol; Gerilite Elixir (Vitarine) and Geritol Liquid (J.B. Williams), both containing 12% alcohol. These preparations may be taken in excessive amounts by those seeking health through intake of nutrient supplements. Cough and cold medicines containing alcohol include:

	Alcohol (%)
Deconamine Elixir (Berlex)	10
Romilar III Decongestant Cough Syrup (Block)	20
Hall's Decongestant Cough Formula (Warner Lambert)	22
Comtrex Liquid (Bristol-Myers)	25
Nyquil Nighttime Colds Medicine Liquid (Vicks)	25

Gastrointestinal drugs that are prescribed for patients with irritable bowel syndrome and that are mixtures containing barbiturates, atropine or belladonna, hyoscamine, and other sedative as well as specific anticholinergic drugs are likely to contain alcohol from 19 to 23 percent. Among antidiarrheal drugs, Dia-Quel (Marion) contains 10% alcohol (Facts and Comparisons 1984).

It is of concern that these preparations may be taken by children and by the frail elderly, as well as by alcoholics receiving disulfiram as an alcohol-aversion drug.

ALCOHOL IN PARENTERAL ALIMENTATION FORMULATIONS

Intravenous infusion of alcohol has been used in parenteral alimentation because ethanol provides a high-energy source that is rapidly metabolized (Grant and Todd 1982). Intravenous solutions containing amino acids and electrolytes in an ethanol and sorbitol formulation are used in the United Kingdom and in Europe. Such preparations include Aminofusin L1000 (B.D.H.) and Aminoplex 5 (Geistlich).

Heuckenkamp et al. (1976) studied the effects of a 5% intravenous ethanol-saline infusion in healthy volunteers. The ethanol was well tolerated and more than 98% of the ethanol was utilized as an energy source. However, blood lactate levels rose rapidly after the infusion was started and remained elevated until the infusion was stopped.

In another study of 17 patients with severe nitrogen depletion entering a gastroenterology ward, Aminoplex 5 was given with parenteral vitamins and supplementary potassium and phosphate (Wells and Smits, 1978). Nitrogen balance was restored most efficiently during the ten-day treatment period in those patients who had the highest rates of infusion. The only metabolic complication encountered was a moderate hypophosphatemia.

The adverse effects of giving vitamins intravenously with alcohol have been identified. Acute folacin deficiency with pancytopenia and megaloblastic erythropoiesis has been reported by Wardrop et al. (1975). The folacin deficiency in this case was conditioned by prior folacin depletion and high requirements for the vitamin, by the presence of infection, and by ethanol infusion.

ALCOHOL (ETHANOL) IN INTRAVENOUS DRUG SOLUTIONS

Intravenous nitroglycerine is used in patients with unstable angina, variant angina, and left ventricular failure. It is also used to treat intraoperative and postoperative hypertension in patients with ischemic heart disease. The nitroglycerine is formulated with ethanol. For every milligram of nitroglycerine given, 0.06 to 0.14 ml of ethanol are infused. Shook et al. (1984) have reported that acute alcohol intoxication may occur when this formulation of nitroglycerine is administered. The two patients who they reported with this complication were in their 70s and both improved when the IV solution of nitroglycerine was discontinued. The authors comment that the ethanol may be cardiotoxic. They also note that several other parenteral drug formula-

TABLE 4.1 DRUG–ALCOHOL AND DRUG–FOOD INCOMPATIBILITY
REACTANTS AND REACTIONS

Reactants		Reactions
Isoniazid	Histamine	Flushing
Phenelzine and	Tyramine in	Hypertensive attacks
other MAOI	food or wine	
drugs		
Disulfiram	Alcohol	Acetaldehyde reaction
Chlorpropamide	Alcohol	Flushing (CPAF)*

*CPAF = acronym for chlorpropamide alcohol flush reactions.

tions including diazepam, trimethoprim-sulfamethoxazole, digoxin, and phenytoin contain ethanol, but since they are not usually administered as a continuous infusion, the risk of intoxication is diminished.

Wernicke's encephalopathy has also been reported as a complication of intravenous nitroglycerine therapy. The encephalopathy was produced by the ethanol and propylene glycol diluents of the drug. Whereas the risk of Wernicke's encephalopathy with alcohol intake and lack of thiamin is well known, the role of propylene glycol has not been extensively discussed. Propylene glycol is metabolized to pyruvate which requires thiamin pyrophosphate for entry into the citric acid cycle as an energy source. Thus, it is possible that small amounts of propylene glycol in the IV solution may have slightly increased the thiamin requirements and contributed to the etiology of Wernicke's encephalopathy (Shorey et al. 1984).

DRUG-INDUCED INHIBITION OF ALCOHOL AND AMINE METABOLISM IN FOOD

Drugs can inhibit the metabolism of alcohol or food components with the resulting release of vasoactive substances and an adverse reaction. Among the reactions are histamine, tyramine, disulfiram, and chlorpropamide-flush reactions. Drugs that induce these reactions include therapeutics such as isoniazid, monamine oxidase inhibitors (MAOI), alcohol-aversion drugs such as tetraethylthiuram disulfide (Disulfiram or Antabuse), and natural food toxins such as the toxin in Coprinus Atramentarius, the inky cap mushroom. Drug–alcohol and drug–food incompatibilities and the reactions they produce are shown in Table 4.1.

Whether or not a reaction occurs and also the severity of such a

TABLE 4.2 FACTORS AFFECTING DRUG-INDUCED
INCOMPATIBILITY REACTIONS

Type of Reaction	When Does It Occur?	How Severe Is It?
Histamine	When fish is eaten by INH user	Depends on histamine content of fish and rate of INH metabolism
Tyramine	When cheese is eaten by MAOI user	Depends on tyramine content of food & BP
Disulfiram	When alcohol is taken with drug	Depends on blood acetaldehyde level
Chlorpropamide flush	When alcohol is taken after drug	Depends whether a Type II diabetic

reaction depend on pharmacogenetic factors as well as on the drug dose, the frequency of drug administration, the foods and beverages consumed, when the drug is being administered, and on the health status of the drug-taker. The health status of the drug-taker not only determines which drug(s) are being taken but also may determine the outcome of the drug-induced incompatibility. These variables and their effects on the occurrence and outcome of incompatibility reactions are shown in Table 4.2.

FLUSHING REACTIONS

Flushing reactions are heterogeneous with respect to site, mode of induction, and triggering agent. Flushing may result from substances or stimuli that act directly on vascular smooth muscle or are mediated by vasomotor nerves. Whereas flushing is usually limited to the face, it may involve the chest and other areas. Flushing occurs in certain disease states, notably in people with rosacea and in those with carcinoid tumors. In rosacea, flushing may be triggered by ingestion of hot food. It is also worsened by vasodilator therapy (Wilkin 1981). Flushing is clinically manifested by histamine, tyramine, disulfiram, and chlorpropamide (CPAF) reactions.

ALCOHOL-ASSOCIATED FLUSH REACTIONS

The flushing that occurs with alcohol ingestion is more likely to occur in people with specific predisposing conditions. Thus, Orientals with a genetically defective acetaldehyde oxidation mechanism flush when they only ingest very small amounts of alcohol. Facial flushing occurs in more than 50% of Orientals after alcohol ingestion, but it is uncommon in Caucasians. Orientals who flush after alcohol may experience other hemodynamic changes such as pallor, hypotension, and vasovagal collapse. The severity of the symptoms is related to the level of acetaldehyde in the blood and to the secondary release of catecholamines (Kupari et al. 1983).

HISTAMINE REACTIONS

Histamine reactions have been reported in patients who ate certain types of fish, including tuna and skipjack, while they were receiving isoniazid for the prevention or treatment of tuberculosis. These histamine reactions were characterized by a severe throbbing headache, facial flushing, and redness and itching of the eyes and palms. Symptoms began from 5 minutes up to 2 hours after a "histamine" meal in individuals who had taken their isoniazid 3 to 4 hours earlier. The symptoms reached a peak about 5 hours after the onset and subsided in about 12 hours (Uragoda 1980).

Scombroid fish, including tuna, bonito, and mackerel, are potent sources of histamine if spoilage has converted the histidine present in the fish (Oehme et al. 1980). Nonscombroid fish have also been reported to cause histamine reactions (Turnbull and Gilbert 1982). In order to prevent histamine reactions, individuals receiving isoniazid should not eat the following fish:

Tuna—fresh or canned
Mackerel—fresh, smoked, or canned
Sardines—canned
Bonito—fresh or canned
Anchovies—fresh or canned
Pilchards—canned

TYRAMINE REACTIONS

Tyramine reactions occur when high-tyramine foods or beverages are consumed by individuals who are receiving drugs that are monamine oxidase inhibitors (MAOI). Tyramine sources include aged cheeses, pickled herring, chicken livers, chianti wines, and some beers. MOAI drugs that have been reported to evoke tyramine reactions consist of the antidepressants phenelzine (Nardil), isocarboxyazid (Marplan), and

TABLE 4.3 FOODS AND BEVERAGES WHICH POSE A SIGNIFICANT
HAZARD TO PATIENTS ON MAOI DRUGS, WITH THEIR
APPROXIMATE TYRAMINE CONTENT

Food or Beverage	Tyramine Content (μg/gm or μg/ml)
Cheese	
Brie	180
Camembert	86
Cheddar, N.Y. State	1416
Gruyere	516
Stilton	466
Fish	
Pickled herring	3030
Alcoholic beverages	
Chianti	25

Source: Overton and Lukert 1977.

tranylcypromine (Parnate); they are prescribed for patients with psychotic depression as well as for those with phobic anxiety. These drugs function by elevating levels of norepinephrine and serotonin in the central nervous system and potentiate the vasopressor effects of simple phenylethylamines, such as tyramine, which act by releasing catecholamines. Other MAOI drugs that can cause tyramine reactions include isoniazid and procarbazine, used in the treatment of Hodgkin's disease (Blackwell and Mabbitt 1965; Blackwell et al. 1965; Asatoor et al. 1963; Roe 1978).

Tyramine reactions are characterized by hypertension of brief duration, headaches, palpitations, nausea, and vomiting. Major cerebrovascular accidents have been reported with massive intake of tyramine. The severity of the attack has also been related to drug dosage. Foods containing dopamine or serotonin may also produce these reactions.

Foods and beverages that pose a significant hazard to patients on MAOI drugs are listed in Table 4.3 together with their approximate tyramine content (Overton and Lukert 1977).

Yeast extracts that contain tyramine and broad beans that contain dopamine have also been reported to cause acute hypertensive reactions in MAOI users. The interaction between the antidepressant drug tranylcypromine and cheddar cheese has actually been used to treat postural hypotension (Diamond et al. 1969). But this is considered a hazardous procedure, in that the tyramine content of cheddar cheese is variable and an untoward response could well be elicited.

DISULFIRAM REACTIONS

Tetraethylthiuram disulfide (Disulfiram or Antabuse) is used extensively as an oral drug or a subcutaneous implant in the management of alcoholism. It is a potent and useful alcohol-aversion drug that produces a reaction when alcohol is consumed even in very small amounts. The disulfiram reaction is characterized by flushing of the face, neck, and upper chest; a throbbing headache, nausea, faintness, and chest or abdominal pain may occur (Morgan and Cagan 1974) even up to 10 days after drug administration ceases.

Disulfiram-like reactions occur when alcoholic beverages are consumed with certain other drugs, including griseofulvin, metronidazole, quinacrine, and tolazoline. Citrated calcium carbimide, which has been used in Canada as an alcohol-aversion drug, produces effects that are identical to that of disulfiram (Roe 1979; Seixas 1975).

It has been pointed out by Faiman (1979) that disulfiram is an inhibitor of numerous enzymes and that the drug can therefore affect carbohydrate metabolism, mitochondrial oxidations, neurotransmission, and drug metabolism. However, the inhibitory effect of disulfiram on aldehyde dehydrogenase with consequent impairment in the metabolism of acetaldehyde after alcohol ingestion is most important. Indeed, most of the clinical effects of administration of alcohol after disulfiram are similar to those induced by administration of acetaldehyde.

Toxic effects of disulfiram include hypersensitivity reactions with development of dermatitis, cardiac arrhythmias, myocardial infarction, optic neuritis, peripheral neuropathy, seizures, and psychotic reactions (Faiman 1979).

Ingestion of the inky cap mushroom (Coprinus atramentarius) prior to the consumption of beer or other alcoholic beverages will elicit a disulfiram-like reaction. It has been postulated that a metabolite of coprine, 1-aminocyclopropanol, induces the effect because, like disulfiram, it is an inhibitor of aldehyde dehydrogenase and therefore would allow build-up of acetaldehyde with the characteristic reaction (Tottmar et al. 1977).

CHLORPROPAMIDE-ALCOHOL FLUSH REACTIONS

Chlorpropamide–alcohol flushing (CPAF) was described by Leslie et al. (1979). These investigators were of the opinion that CPAF occurred in mild diabetes of the dominantly inherited, noninsulin-dependent,

"Mason type" that carries a relatively good prognosis and is seldom complicated by retinopathy. This group also reported on a test for CPAF in which they gave 40 ml of sherry 12 hours after a single dose of 250 mg chlorpropamide. They found a high correlation between flushing in this single challenge test and a history of flushing when alcoholic beverages were taken after chlorpropamide (33 out of 35 patients). They also found a positive CPAF reaction in 51% of noninsulin-dependent diabetics, in contrast to 10% of insulin-dependent diabetics and 10% of nondiabetic controls. Their experience with this test was that it could be used to identify some "Mason-type" diabetics who are not only less susceptible to retinopathy, but also are less likely to develop peripheral vascular disease or ischemic heart disease (Leslie et al. 1979; Barnett and Pyke 1980).

Relative protection against late complications of diabetes was also found by Jerntorp and Almer (1981) in CPAF-positive diabetics. In particular, they found that signs of peripheral vascular disease and peripheral neuropathy were less common in CPAF-positive patients. Their findings have not been fully confirmed by other investigators.

Among 50 diet-treated, noninsulin-dependent diabetics who were screened for CPAF by De Silva et al. (1981), 24% reported a subjective flush. But of these, 18% also flushed when a placebo was given instead of the chlorpropamide. In a control group of 21 nondiabetics, two showed the CPAF reaction.

In a study by Groop et al. (1984), 160 nonketotic diabetics aged 35 to 70 years were given the CPAF test. The authors confirmed previous observations that the CPAF-positive reaction had a higher incidence in diabetics with a family history of the disease. HLA typing was performed and the CPAF-positive group was found to have a significant increase in HLA-A2 and a lower frequency of HLA-B7. However, it was not possible to demonstrate a difference in the incidence of diabetic complications between the CPAF-positive and -negative groups.

Suggestions have been made to explain the variation in results with CPAF patients, including the crudeness of the sherry test and the effect of drug duration. The etiology of the flush reaction has also been investigated. Medbak et al. (1981), on the basis of small-scale testing, concluded that endogenous opiates may be implicated.

Barnett et al. (1981) found that CPAF-positive diabetics have higher levels of acetaldehyde in their blood after testing. They suggest on the basis of these findings that CPAF-positive patients may have a genetically determined alcohol dehydrogenase isoenzyme that is very susceptible to chlorpropamide inhibition. A further proposal with wide support is that the CPAF reaction is prostaglandin-mediated; it is

**TABLE 4.4 DRUG–ALCOHOL INTERACTIONS AND THEIR
ADVERSE OUTCOMES**

Drug Group	Alcohol Effect
Barbiturates	Additive or synergistic CNS depressant effect (Deitrich and Petersen 1979)
Coronary Vasodilators (nitroglycerin)	Hypotension can result from use of alcohol and nitroglycerine (Shafer 1971)
Non-steroid Anti-inflammatory drugs (aspirin, indomethacin)	GI bleeding due to salicylates or indomethacin is enhanced by alcohol (Dobbing 1969)
Antihistamines	Alcohol increases sedative effect of antihistamines because capsules are rapidly dissolved (Parker 1970)
Antigout Drugs (Colchicine)	Alcohol increases colchicine-induced lactose intolerance (Holtzapple and Schwartz 1984)
Anticonvulsants (phenytoin), diuretics (furosemide)	Alcohol excess increases the risk of phenytoin-induced folate deficiency and diuretic-induced magnesium deficiency (Roe 1985)
Oral hypoglycemic agents (chlorpropamide) and alcohol-aversion drugs (Disulfiram)	Alcohol ingestion triggers drug-related flush reactions (Wilkin 1981)
Cytotoxic agents (Methotrexate)	Alcohol enhances the hepatotoxicity of methotrexate (Zachariae et al. 1980)
Gaseous anesthetics	Alcohol abuse increases tolerance for anesthetics (Deitrich and Petersen 1979)
Tricyclic antidepressants (Imipramine)	Dyskinetic effects of antidepressants are increased in alcoholics (Byck 1975)

suppressed by aspirin and by indomethacin (Stakosch et al. 1980; Barnett et al. 1980).

ADVERSE DRUG REACTIONS DUE TO THE ADDITIVE OR SYNERGISTIC EFFECTS OF ALCOHOL (ETHANOL) ON OTHER DRUG EFFECTS OR REACTIONS

1 The depressant effect of alcohol on the central nervous system (CNS) potentiates the depressant effects of other drugs.
2 Vasodilator and related hypotensive effects of alcohol potentiate similar effects induced by other drugs.

3   Locally irritant effects of alcohol on the gastrointestinal tract with development of gastritis potentiates the gastritis induced by other drugs.
4   Alcohol dissolves capsules and can increase the rate of dissolution and absorption of drugs given concurrently in capsule form.
5   Alcohol potentiates the lactase inhibitory effects of other drugs, causing lactose intolerance.
6   Alcohol potentiates the adverse nutritional effects of other drugs that cause anorexia, maldigestion, malabsorption, vitamin antagonism, and mineral wasting.
7   Alcohol ingestion triggers acute incompatibility reactions related to intake of other drugs.
8   Tissue toxicity of foreign compounds, including drugs, is potentiated by alcohol-related damage to the same tissues.
9   Heavy chronic ingestion of alcohol increases tolerance for other drugs.
10 Risk of specific adverse drug reactions is increased during periods of heavy alcohol consumption (Roe 1979).

Examples of these drug–alcohol interactions are shown in Table 4.4.

REFERENCES

ASATOOR, A.M., A.J. LEVI, and M.D. MILNE. 1963. Tranylcypromine and cheese. *Lancet* 2: 733–34.

BAILEY, D. 1975. The alcohol content of some commonly prescribed medicines. *J. Alcoholism* 10: 67–72.

BARNETT, A.H., C. GONZALEZ-AUVERT, D.A. PYKE, J.B. SAUNDERS, R. WILLIAMS, C.J. DICKENSON, and M.D. RAWLINS. 1981. Blood concentrations of acetaldehyde during chlorpropamide-alcohol flush. *Brit. Med. J.* 283: 939–41.

BARNETT, A.H. and D.A. PYKE. 1980. Chlorpropamide alcohol flushing and large vessel disease in non–insulin dependent diabetes. *Brit. Med. J.* 281: 261–62.

BARNETT, A.H., A.J. SPILIOPOULOS, and D.A. PYKE. 1980. Blockade of chlorpropamide-alcohol flushing by indomethacin suggests an association between prostaglandins and diabetic complications. *Lancet* 2: 164–66.

BLACKWELL, B. and L.A. MABBITT. 1965. Tyramine in cheese related to hypertensive crises after monamine oxidase inhibition. *Lancet* 1: 940–43.

BLACKWELL, B., L. MARLEY, and L.A. MABBITT. 1965. Effects of yeast extract after monamine oxidase inhibition. *Lancet* 1: 944.

BYCK, R. 1975. Drugs and the treatment of psychiatric disorders. In *The Pharmacological Basis of Therapeutics*, 5th Ed. ed. L.S. Goodman and A. Gilman. 152–200. New York: Macmillan.

DEITRICH, R.A. and D.R. PETERSEN. 1979. Interaction of ethanol with drugs. In *Biochemistry and Pharmacology of Ethanol*, vol. 2. ed. E. Majchrowicz and E.P. Noble. 283–302. New York: Plenum.

DESILVA, N.E., W.M.G. TUNBRIDGE, and K.G.M.M. ALBERTI. 1981. Low incidence of chlorpropamide-alcohol flushing in diet-treated, non-insulin-dependent diabetes. *Lancet* 1: 128–31.

DIAMOND, M.A., R.H. MURRAY, and P. SCHMID. 1969. Treatment of idiopathic postural hypotension with oral tyramine (TY) and monamine oxidase inhibitor (MI). *Clin. Res.* 17: 237.

DOBBING, J. 1969. Faecal blood loss after sodium acetylsalicylate taken with alcohol. Letter. *Lancet* 1: 527.

Facts and Comparisons Drug Information. 1984. St. Louis: Facts and Comparisons, Inc.

FAIMAN, M.D. 1979. Biochemical pharmacology of disulfiram. In *Biochemistry and*

Pharmacology of Ethanol, vol. 2 ed. E. Majchrowicz and E.P. Noble. 325–48. New York: Plenum.

GRANT, A. and E. TODD. 1982. *Enteral and Parenteral Nutrition, A Clinical Handbook.* 148. Oxford: Blackwell.

GROOP, L., S. KOSKIMIES, and E-M. TOLPANNEN. 1984. Characterization of patients with chlorpropamide-alcohol flush. *Acta Med. Scand.* 215: 141–49.

HEUKENKAMP, P-U, U. SPRANDEL, and E.W. LIEBHARDT. 1976. Studies concerning ethanol as a nutrient for intravenous alimentation in man. *Nutr. Metab.* 21 Suppl. 1: 121–24.

HOLTZAPPLE, P.G. and S.E. SCHWARTZ. 1984. Drug-induced maldigestion and malabsorption. In *Drugs and Nutrients: The Interactive Effects.* ed. D.A. Roe and T.C. Campbell. 475–85. New York: Marcel Dekker.

JERNTORP, P. and L-O. ALMER. 1981. Chlorpropamide-alcohol flushing in relation to macroangiopathy and peripheral neuropathy in non–insulin dependent diabetes. *Acta Med. Scand. Suppl.* 656: 33–36.

KUPARI, M., P. ERIKSSON, J. HEIKKILA, and R. YLIKAHRI. 1983. Alcohol and the heart. *Acta Med. Scand.* 213: 91–98.

LESLIE, R.D.G., A.H. BARNETT, and D.A. PYKE. 1979. Chlorpropamide alcohol flushing and diabetic retinopathy. *Lancet* 1: 997–99.

LESLIE, R.D.G. and D.A. PYKE. 1978. Chlorpropamide alcohol flushing: a dominantly inherited trait associated with diabetes. *Brit. Med. J.* 2: 1519–21.

MEDBAK, S., J.A.H. WASS, V. CLEMENT-JONES, E.D. COOK, S.A. BOWCOCK, A.G. CUDWORTH, and L.H. REES. 1981. Chlorpropamide alcohol flush and circulating met-enkephalin: a positive link. *Brit. Med. J.* 283: 937–39.

MORGAN, R. and E.J. CAGAN. 1974. Acute alcohol intoxication, the disulfiram reaction and methyl alcohol intoxication. In *The Biology of Alcoholism,* vol. 3. ed. B. Kissin and H. Begleiter. 163–89. New York: Plenum.

OEHME, F.W., J.F. BROWN, and M.E. FOWLER. 1980. Toxins of animal origin. In *Caserett and Doull's Toxicology: The Basic Science of Poisons,* 2nd ed. ed. J. Doull, C.D. Klaassen, and M.O. Amdur. 563. New York: Macmillan.

OVERTON, M. and B. LUKERT. 1977. *Clinical Nutrition: A Physiological Approach.* 161. Chicago: Yearbook Med. Publ. Inc.

PARKER, W.J. 1970. Clinically significant alcohol drug interactions. *J. Am. Pharm. Assoc.* 10: 664.

ROE, D.A. 1978. Diet-drug interactions and incompatibilities. In *Nutrition and Drug Interrelations.* ed. J.N. Hathcock and J. Coon. 319–45. New York: Academic Press.

————. 1979. *Alcohol and the Diet.* 119–48, 208–22. Westport, CT: AVI.

————. 1985. *Drug-Induced Nutritional Deficiencies,* 2nd ed. Westport, CT: AVI.

SEIXAS, F.A. 1975. Alcohol and its drug interactions. *Ann. Intern. Med.* 83: 86–92.

SHAFER, N. 1971. Hypotension due to nitroglycerin combined with alcohol. *New Eng. J. Med.* 273: 1169.

SHOOK, T.L., J.M. KIRSHENBAUM, R.F. HUNDLEY, J.M. SHOREY, and G.A. LA-

MAS. 1984. Ethanol intoxication complicating intravenous nitroglycerin therapy. *Ann. Intern. Med.* 101: 498–99.

SHOREY, J., N. SHARDWAJ, and J. LOSCALZO. 1984. Acute Wernicke's encephalopathy after intravenous infusion with high-dose nitroglycerin. *Ann. Intern. Med.* 101: 500.

STAKOSCH, C.R., D.B. JEFFREYS, and H. KEEN. 1980. Blockade of chlorpropamide alcohol flush by aspirin. *Lancet* 1: 394–96.

TOTTMAR, O., H. MARCHNER, and P. LINDBERG. 1977. *Alcohol and Aldehyde Metabolizing Systems,* vol. 2. ed. R.G. Thurman, J.R. Williamson, H.R. Drott, and B. Chance. 203–12. New York: Academic Press.

TURNBULL, P.C.B. and R.J. GILBERT. 1982. Fish and shellfish poisoning in Britain. In *Adverse Effects of Foods.* ed. E.F.P. Jelliffe and D.B. Jelliffe. 297–306. New York: Plenum Press.

URAGODA, K.B. 1980. Histamine poisoning in tuberculous patients after ingestion of tuna fish. *Am. Rev. Resp. Dis.* 121: 157–59.

WARDROP, C.A.J., G.B. TENNANT, R.V. HEATLEY, and L.E. HUGHES. 1975. Acute folate deficiency in surgical patients on amino acid/ethanol intravenous nutrition. *Lancet* 2: 640–42.

WELLS, F.E. and B.J. SMITS. 1978. Utilization and metabolic effects of a solution of amino acids, sorbitol and ethanol in parenteral nutrition. *Am. J. Clin. Nutr.* 31: 442–50.

WILKIN, J.K. 1981. Flushing reactions: consequences and mechanisms. *Ann. Intern. Med.* 95: 468–76.

ZACHARIAE, H., K. KRAGBULLE, and H. SOGAARD. 1980. Methotrexate induced liver cirrhosis. *Brit. J. Dermat.* 102: 407–12.

Chapter 5

INTOLERANCE TO INTENTIONAL ADDITIVES IN FOODS AND DRUGS

TARTRAZINE (FD&C YELLOW #5)

The yellow dye tartrazine, which is used as a colorant in foods, beverages, and drugs, frequently induces pseudoallergic reactions. Ingestion of this dye is a common cause of acute urticaria and may exacerbate chronic urticaria (Juhlin et al. 1972). According to Warin and Smith (1982), the incidence of reactions to tartrazine in patients with chronic urticaria is about 10%. However, once the urticaria has been clear for some time, patients no longer react to tartrazine.

In some individuals, tartrazine can also provoke an attack of angioedema (Simon 1984). It causes bronchoconstriction in a high percentage of aspirin-sensitive asthmatics (Anderson 1984), and it can also initiate asthma attacks (Settipane 1983).

Reactions to tartrazine appear to be dose-dependent. Clinical investigators have advocated oral provocation tests, with increasing doses of the dye, as a routine measure to test for respiratory or cutaneous reactions (Genton et al. 1985). However, such testing procedures should be conducted only where epinephrine and corticosteroids are available to treat patients who develop positive reactions.

Common commercial food sources of tartrazine include foods that look yellow: frozen baked goods, snacking cakes, cookies, cheese crackers, cake mixes, ready-to-eat breakfast cereals, pudding mixes, ice cream and ice milk, sherbet, gravy mixes, bottled sauces, mustard relish, seasonings, dried potatoes, meat and fish extenders, frozen vegetables in cheese sauce, and fried onion rings. Tartrazine may also be found in powdered formula foods (Freydberg and Gortner 1982).

Prescription and over-the-counter drug sources of tartrazine include antibiotics, antidepressants, antihypertensive drugs, other drugs used in the treatment of cardiovascular disease, tranquilizers, laxatives, and vitamin and/or mineral supplements. Tartrazine has been used as a colorant for product identification by dose; for any drug available in several dosages, only specific forms may contain tartrazine. In recent years, alternate means of product dose identification have been used by pharmaceutical companies. For example, all drugs marketed by Ayerst are tartrazine-free.

Common prescription drugs containing tartrazine are listed in *Facts and Comparisons* (1985, 1986). That information may not be inclusive, and health care providers are urged to maintain an updated list that covers the drugs used in their practice and to check if the drugs listed still contain tartrazine.

SALICYLATES

Salicylates occur naturally in foods and are present in high concentrations in herbs and spices (Swain et al. 1985). Two salicylates, namely acetyl salicylate (or aspirin) and methyl salicylate (or oil of wintergreen) cause acute reactions, of which urticaria is perhaps the most common.

Aspirin intolerance may be manifested as asthma, severe rhinitis, urticaria, angioedema, and occasionally as anaphylactic shock. Individuals with aspirin intolerance may also be sensitive to other drugs, such as indomethacin, mefenamic acid, ibuprofen, and phenylbutazone, as well as to tartrazine and sodium benzoate which are used as food and drug additives. Salicylate-free diets have been devised, the claim being made that such diets may be of use in controlling these conditions (Doeglas 1977). However, the prescribed diets were also free of food additives, such as tartrazine, which can provoke aspirin intolerance (Noid et al. 1974).

Most of the drugs to which aspirin-sensitive people cross-react are prostaglandin inhibitors, but tartrazine and sodium benzoate are not in this category (Settipane 1983). Indeed, at present it is not clear, either from the chemical structure or from the pharmacological effect, what makes a "sensitizing" substance. Sodium salicylate does not have this effect. However, substances that do provoke reactions may cause mast cell degranulation and histamine release (Kallos and Schlumberger 1980).

SULFITES

Acute reactions to sulfites have been increasingly reported in the past five years. Cutaneous reactions include flushing, itching, tingling, and angioedema. Respiratory and general systemic reactions include bronchospasm and hypotension. Sudden death has occurred after exposure

(Settipane 1984). The onset of symptoms is rapid after exposure, but the actual time of appearance depends on the form in which the sulfite is presented. Sulfites may be ingested with food or when sulfite-containing drugs are given by mouth, by inhalation, topically, or by injection or infusion (Keller 1971; Settipane 1984).

Sulfites are antioxidants used in the preservation of foods and drugs. Sulfites on the GRAS (Generally Recognized as Safe) list of the Food and Drug Administration are sulfur dioxide, sodium sulfite, sodium and potassium bisulfite, and sodium metabisulfite. In aqueous solution, sulfites generate SO_2, which may be responsible for reactions to inhaled drugs containing sulfites. Bisulfites are used as antioxidants in a large variety of drugs, including many which are used in the treatment of asthma. Bisulfites are additives in local anesthetics, antibiotics, antidepressants, and ophthalmic solutions as well as parenteral amino acid infusions.

In restaurants, sulfites have been sprayed onto seafood, potatoes, salad greens, avocados, and other produce in salad bars, in order to maintain the appearance of freshness (in 1986 the Food and Drug Administration banned this practice). Sulfites are used in processed foods such as dried potatoes, in wines, beer, cider, and some fruit drinks (FDA 1984; Jamieson et al. 1985; Oberste 1984). The average U.S. diet contains 2 to 3 mg of sulfite per day.

The principal victims of sulfite reactions are asthmatics. Asthmatics are particularly vulnerable, both because of their predisposition to react and because of their enhanced risk of exposure to drugs that contain sulfites. Sulfite sensitivity may also result from a sulfite oxidase deficiency. This genetic defect was demonstrated by Jacobsen et al. (1984), who also found that cyanocobalamin (vitamin B_{12}) can act as an extracellular catalyst for the oxidation of sulfites in deficient individuals and that it can afford them protection against sulfite reactions.

Atropine and cromolyn have been used to protect sulfite-sensitive people from reactions. Inhibition of sulfite reactions by atropine has been used to support the hypothesis that the reactions are due to parasympathetic stimulation (Freedman 1977). On the other hand, the response to cromolyn indicates that the reactions are related to histamine release, which is inhibited by this drug.

MONOSODIUM GLUTAMATE

Monosodium glutamate (MSG) has long been used as a flavor enhancer in Oriental foods and is the active ingredient in "Accent." Although the FDA has placed MSG on the GRAS list, it may not be safe for certain individuals to consume foods containing MSG. MSG has been implicated as the cause of the so-called Chinese Restaurant Syndrome (CRS), which consists of facial flushing, headache, discomfort in the neck and shoulders, and faintness that occurs in some individuals after eating Chinese food (Monsereenusorn 1982; Gore 1982). Whereas CRS used to occur most frequently when susceptible individuals ate Chinese food in a restaurant, the present availability of Chinese "take-out" foods, convenience foods, and "Oriental" rice has led to complaints of CRS by people dining at home.

MSG may be neurotoxic. Experiments have shown that MSG can cause brain lesions in neonatal mice and monkeys (Olney 1969; Olney and Sharpe 1969). While it has not been demonstrated that MSG is neurotoxic to human subjects, there is still some fear that MSG may injure the brain of the young infant and it is therefore considered unwise to add it to infant foods. Reasons for susceptibility to MSG are not well defined.

According to Kwok (1968), reactions to MSG might include the toxic effects of intake of large amounts of glutamate or sodium. Kenney (1979, 1980) proposed that MSG caused reactions in people who have gastroesophageal reflux, and he also carried out studies which suggested that perhaps CRS was due to esophageal stimulation. Lessof (1985) has offered the alternative hypothesis that CRS is a conditioned toxic reaction that depends on genetic predisposition.

CAFFEINE

Caffeine is widely consumed as an additive to beverages, as a component of prescription and over-the-counter drugs, and as an important pharmacologic substance in coffee, tea, cocoa, and chocolate. There has been concern both by the public and by the FDA that since caffeine is added to certain carbonated soft drinks, children may exhibit adverse behavioral effects (Miller and Harris 1983). Positive effects of caffeine include decreased fatigue, greater alertness, and longer attention span in hyperactive youngsters. Negative effects include insomnia, nervousness, cardiac arrhythmias, and headaches on caffeine withdrawal. The occurrence of arrhythmias is related to dose and time of administration (Rall 1980; Dobmeyer et al. 1983). Caffeine has been identified as having potential fetal toxicity (Collins et al. 1982). Information on caffeine-containing beverages, foods, and drugs is shown in Tables 5.1 and 5.2.

TABLE 5.1 CAFFEINE CONTENT OF FOOD PRODUCTS

Product	Vol. or Wt.	Caffeine Content Range[1]	Caffeine Content Average	Reference
Roasted & ground coffee (perc.)	5 oz	64–124 mg	83 mg	Burg (1975)
Instant coffee	5 oz	40–108 mg	59 mg	Burg (1975)
Roasted & ground coffee (decaff)	5 oz	2–5 mg	3 mg	Burg (1975)
Roasted & ground coffee (drip)	5 oz	—	112 mg	Gilbert (1981)
Tea	5 oz	8–91 mg	27 mg	Gilbert (1981)
Bagged tea	5 oz	—	42 mg	Burg (1975)
Leaf tea	5 oz	30–48 mg	41 mg	Burg (1975)
Instant tea	5 oz	24–31 mg	28 mg	Burg (1975)
Cocoa—African	5 oz	—	6 mg	Burg (1975)
Cocoa—South American	5 oz	—	42 mg	Burg (1975)
Cocoa from mix	5 oz	—	6 mg	IFT (1981)
Baking chocolate	1 oz	—	35 mg	IFT (1983)
Baking chocolate	1 oz	18–118 mg	60 mg	Zoumas et al. (1980)
Milk chocolate bar	1 oz	1–15 mg	6 mg	Zoumas et al. (1980)
Chocolate bar	1 oz	—	20 mg	Gilbert (1981)
Chocolate milk	8 oz	2–7 mg	5 mg	Zoumas et al. (1980)
Sugar-free Mr. Pibb	12 oz	—	58.8 mg	Nat. Soft Drink Assoc. (1982)
Mountain Dew	12 oz	—	54.0 mg	Nat. Soft Drink Assoc.
Mello Yello	12 oz	—	52.8 mg	Nat. Soft Drink Assoc.
Tab	12 oz	—	46.8 mg	Nat. Soft Drink Assoc.
Coca-Cola	12 oz	—	45.6 mg	Nat. Soft Drink Assoc.
Diet Coke	12 oz	—	45.6 mg	Nat. Soft Drink Assoc.
Shasta Cola	12 oz	—	44.4 mg	Nat. Soft Drink Assoc.
Shasta Cherry Cola	12 oz	—	44.4 mg	Nat. Soft Drink Assoc.
Shasta Diet Cherry Cola	12 oz	—	44.4 mg	Nat. Soft Drink Assoc.
Mr. Pibb	12 oz	—	40.8 mg	Nat. Soft Drink Assoc.

(continued)

TABLE 5.1 CAFFEINE CONTENT OF FOOD PRODUCTS (CONT.)

Product	Vol. or Wt.	Caffeine Content Range[1]	Caffeine Content Average	Reference
Dr. Pepper	12 oz	—	39.6 mg	Nat. Soft Drink Assoc.
Sugar-Free Dr. Pepper	12 oz	—	38.4 mg	Nat. Soft Drink Assoc.
Pepsi-Cola	12 oz	—	38.4 mg	Nat. Soft Drink Assoc.
Diet Pepsi	12 oz	—	36.0 mg	Nat. Soft Drink Assoc.
Pepsi Light	12 oz	—	36.0 mg	Nat. Soft Drink Assoc.
RC Cola	12 oz	—	36.0 mg	Nat. Soft Drink Assoc.
Diet Rite[2]	12 oz	—	36.0 mg	Nat. Soft Drink Assoc.
Canada Dry Jamaica Cola	12 oz	—	30.0 mg	Nat. Soft Drink Assoc.
Canada Dry Diet Cola	12 oz	—	1.2 mg	Nat. Soft Drink Assoc.
Club Soda	12 oz	—		Nat. Soft Drink Assoc.

[1]Representative values have been selected for caffeine-containing beverages and these values have been used by Barone and Roberts (1984) to compute the caffeine consumption of various subgroups of the population. The values used by these authors are as follows:

Coffee—ground and roasted	85 mg/5 oz cup.
—instant	60 mg/5 oz cup.
—decaffeinated	3 mg/5 oz cup.
Tea—leaf or bag	40 mg/5 oz cup
—instant	30 mg/5 oz cup
Cola—except caffeine-free	36 mg/6 oz glass
Cocoa or hot chocolate	4 mg/5 oz cup
Chocolate milk	5 mg/8 oz glass.

[2]The major soft drink companies have recently introduced caffeine-free colas and other sodas. These caffeine-free products include Coke, Diet Rite Cola, A&W Root Beer, Seagram's Ginger Ale, 7-Up, Sunkist, and Snapple sodas.

Sources: Barone, J.J. and H. Roberts. 1984. Human consumption of caffeine. In *Caffeine: Perspectives from Recent Research*. ed. P.B. Dews. pp. 59–73. New York: Springer-Verlag.
Burg, A.W. 1975. How much caffeine in the cup? *Tea Coffee Trade J*. 147:40–42.
Caffeine. 1983. A scientific status summary by the Institute of Food Technologists' Expert Panel on Food Safety and Nutrition.
Gilbert, R.M. 1981. Caffeine: overview and anthology. In *Nutrition and Behavior*. ed. S.A. Miller. pp. 145–66. Philadelphia, PA: Franklin Institute Press.
Zoumas, B.L., W.R. Kreiser, and R. A. Martin. 1980. Theobromine and caffeine content of chocolate products. *J. Food Sci*. 45:314–16.
National Soft Drink Association. 1982. What's in soft drinks? 2nd ed. Washington.

TABLE 5.2 DRUGS CONTAINING CAFFEINE

Brand	Dose	Company
Amaphen	40 mg	Trimen
Anacin Analgesic	32 mg	Whitehall
Anoquan	40 mg	Mallard
APAP Fortified Tabs	65 mg	Rugby
Buff-A-Comp	40 mg	Mayrand
Buffets II Tabs	32.4 mg	Bowman
Butal Compound	40 mg	Cord Labs
Cafergot	100 mg	Sandoz
Cafergot P-B	100 mg	Sandoz
Cafergot P-B Suppos.	100 mg	Sandoz
Cafetrate	100 mg	Schein
Caffedrine	200 mg	Thompson
Caffeine	200 mg	Rugby
Caffeine & Na. Benzoate	500 mg	Lilly
Citra Caps	30 mg	Boyle
Citrated Caffeine	65 mg	Lilly
Coryban-D Caps	30 mg	Pfipharmecs
Dexadar	200 mg	Republic
Dexadar-Plus	200 mg	Republic
Dex-A-Diet, original	200 mg	O'Connor
Dexatrim	200 mg	Thompson
Dexitac	250 mg	Republic
Dietac	200 mg	Menley & James
Dristan Tabs	16.2 mg	Whitehall
Emprazil-C Tabs	30 mg	Burroughs-Wellcome
Endelor	40 mg	Keene
Ercaf	100 mg	Geneva Generics
Ercatab	100 mg	Cord
Ergo-Caff	100 mg	Rugby
Esgic	40 mg	Gilbert
Excedrin	65 mg	Bristol-Myers
Femcapps	32.4 mg	Otis Clapp
Fiorinal	40 mg	Sandoz
G-1	40 mg	Hauck
Isollyl (improved)	40 mg	Rugby
Kolephrin Caps	65 mg	Pfeiffer
Lanorinal	40 mg	Lannett
Marnal	40 mg	Vortech
Medigesic Plus	40 mg	U.S. Chemical Mkt

(continued)

TABLE 5.2 DRUGS CONTAINING CAFFEINE (CONT.)

Brand	Dose	Company
Midol Caplets	32.4 mg	Glenbrook
Migral Tablets	50 mg	Burroughs-Wellcome
No Doz	100 mg	Bristol-Myers
Percodan	32 mg	Endo
Percodan-Demi	32 mg	Endo
Prolamine	140 mg	Thompson
Protension	40 mg	Dwyer
Quick Pep	150 mg	Thompson
Repan	40 mg	Everett
Resolution 1	140 mg	Lee Pharm
Summit Caps	100 mg	Pfeiffer
Tenstan	40 mg	Halsom
Tirend	100 mg	Norcliff Thayer
Triaminicin Tabs	30 mg	Dorsey
Two-Dyne	40 mg	Hyrex
Vanquish	33 mg	Glenbrook
Vivarin	200 mg	Beecham Products
Westrim C	140 mg	Western Research
Wigraine Suppos	100 mg	Organon
Wigraine Tablets	100 mg	Organon

REFERENCES

ANDERSON, J.A. 1984. Non-immunologically-mediated food sensitivity. *Nutr. Rev.* 42: 109–16.

COLLINS, T.F.X., J.J. WELSH, T.N. BLACK, and D.I. RUGGLES. 1982. Teratogenic potential of caffeine in rats. In *Alternative Dietary Practices and Nutritional Abuses in Pregnancy.* 97–107. Proc. Workshop. Nat. Acad. Press.

DOBMEYER, D.J., R.A. STINE, C.V. LEIER, R. GREENBERG, and S.F. SCHAAL. 1983. The arrhymogenic effects of caffeine in human beings. *New Eng. J. Med.* 308: 814–16.

DOEGLAS, H.M. 1977. Dietary treatment of patients with chronic urticaria and intolerance to aspirin and food additives. *Dermatologica* 154: 308.

Facts and Comparisons, 1985.

FDA DRUG BULLETIN. 1984. Sulfite update. *FDA Drug Bull.* 14: 24.

FREEDMAN, B.J. 1977. Asthma induced by sulfur dioxide, benzoate and tartrazine containing orange drinks. *Clin. Allergy* 7: 407–15.

FREYBERG, N. and W.A. GORTNER. 1982. *The Food Additives Book.* New York: Bantam.

GENTON, C., C. FREI, and A. PECOUD. 1985. Value of oral provocation tests to aspirin and food additives in the routine investigation of asthma and chronic urticaria. *J. Allergy Clin. Immunol.* 76: 40–45.

GORE, M. 1982. The Chinese Restaurant Syndrome. In *Adverse Effects of Foods.* ed. E.F.P. Jelliffe and D.B. Jelliffe. 211–23. New York: Plenum.

JACOBSEN, D.W., R.A. SIMON, and M. SINGH. 1984. Sulfite oxidase deficiency and cobalamin protection in sulfite-sensitive asthmatics (SSA). Abstract. *J. Allergy Clin. Immunol.* 73: 135.

JAMIESON, D.M., M.J. GUILL, and B.B. WRAY. 1985. Metabisulfite sensitivity: Case report and literature review. *Ann. Allergy* 54: 115–21.

JUHLIN, L., G. MICHAELSSON, and O. ZETTERSTROM. 1972. Urticaria and asthma induced by food-and-drug additives in patients with aspirin hypersensitivity. *J. Allergy Clin. Immunol.* 50: 92–98.

KALLOS, P. and H.D. SCHLUMBERGER. 1980. The pathomechanism of acetylsalicylic acid intolerance. A hypothesis. *Med. Hypotheses* 6: 487–90.

KELLER, D.F. 1971. G-6-PD Deficiency. London: CRC, Butterworth's.

KENNEY, R.A. 1979. Placebo controlled studies of human reaction to oral monosodium

L-glutamate. In *Glutamatic Acid: Advances in Biochemistry and Physiology*. ed. L.J. Filer et al. 363–73. New York: Raven Press.

———. 1980. Chinese Restaurant syndrome. *Lancet* 1: 311–12.

KWOK, R.H.M. 1968. Chinese Restaurant Syndrome. *New Eng. J. Med.* 278: 796.

LESSOF, M.H. 1985. Food intolerance. *Proc. Nutr. Soc.* 44: 121–25.

MILLER, S.A. and J.E. HARRIS. 1983. Drugs in the food supply. In *Nutrition and Drugs*. ed. M. Winick. 89–99. New York: Wiley.

MONSEREENUSORN, Y. 1982. Common food additives and spices in Thailand. Toxicological effects. In *Adverse Effects of Foods*. ed. J.F.P. Jelliffe and D.B. Jelliffe. 195–202. New York: Plenum Press.

NOID, H.E., T.W. SCHULTZ, and R.K. WINKELMANN. 1974. Diet plan for patients with salicylate-induced urticaria. *Arch. Dermatol.* 109: 866–69.

OBERSTE, D.M. 1984. Health update on sulfites in foods and drugs. *J. Arkansas Med. Soc.* 81: 130–32.

OLNEY, J.W. 1969. Brain lesions, obesity and other disturbances in mice treated with monosodium glutamate. *Science* 164: 719–21.

OLNEY, J.W. and L.G. SHARPE. 1969. Brain lesions in an infant rhesus monkey treated with monosodium glutamate. *Science* 166: 386–88.

RALL, T.W. 1980. The xanthines. In *The Pharmacological Basis of Therapeutics*, 2nd ed. ed. L.S. Goodman and A. Gilman. 592–607. New York: Macmillan.

SETTIPANE, G.A. 1983. Aspirin and allergic diseases: a review. *Amer. J. Med.* 74: 102–9.

———. 1984. Adverse reactions to sulfites in drugs and foods. *J. Amer. Acad. Dermatol.* 10: 1077–80.

SIMON, R.A. 1984. Adverse reactions to drug additives. *J. Allergy Clin. Immunol.* 74: 623–30.

SWAIN, A.R., S.P. DUTTON, and A.S. TRUSWELL. 1985. Salicylates in foods. *J. Amer. Diet. Assoc.* 85: 950–60.

WARIN, R.P. and R.J. SMITH. 1982. Role of tartrazine in chronic urticaria. *Brit. Med. J.* 284: 1443–44.

Chapter 6

DRUG-INDUCED NUTRITIONAL DEFICIENCIES

Drugs can cause impairment in nutritional status when they produce one or more of the following effects:
1   Anorexia with depression of food intake
2   Malabsorption
3   Vitamin antagonism
4   Mineral depletion
5   Catabolic stress with loss of lean body mass.

A summary list of drugs that cause nutrient depletion and deficiency is found in Table 6.1.

ANTACIDS

Antacids contain calcium, aluminum, magnesium, and sodium as the hydroxides, carbonates, and bicarbonates. Products containing significant amounts of sodium are contraindicated for patients with hypertension or congestive heart failure. In response to demand, the labels of a number of newer antacids indicate low sodium content.

Antacids bring symptomatic relief of indigestion, acute gastritis, peptic esophagitis, gastric hyperacidity, and hiatal hernia. They are used in the treatment of gastric and duodenal ulcers. Aluminum-containing antacids can bind phosphate to control hyperphosphatemia in patients with end-stage renal disease. Antacids are useful to patients receiving pancreatic enzyme preparations in the treatment of maldigestion due to cystic fibrosis or chronic pancreatitis. Calcium-containing antacids can serve as a supplementary calcium source to prevent or delay osteoporosis. The most common reasons for lack of therapeutic efficacy of antacids are insufficient or irregular dosage. Antacids that contain defoamers are claimed to relieve "gas" (Roe 1984; Physicians' Desk Reference 1980).

It was previously accepted by gastroenterologists that if antacids were to be used in the management of peptic ulcer, they should be given one hour and three hours postprandially and at bedtime. Doses during the night have been advocated. High-dosage schedules, or intake of large amounts of antacids at mealtimes, increase the risks of adverse side effects, including drug–drug interactions and drug–nutrient interactions that are of special concern here.

The chemical nature of antacids determines their capacity to affect the absorption of a number of therapeutic drugs. Calcium-, magnesium-, and aluminum-containing antacids interfere with the absorption of tetracycline. Aluminum-containing antacids reduce the bioavailability of digoxin, isoniazid, phenytoin, glucocorticoids, quinidine, and warfarin. Antacids can impair the absorption of anticholinergic drugs and phenothiazines. The excretion of drug bases such as amphetamines and quinidine may be slowed by antacids; increased risk of toxicity may result (Facts and Comparisons 1984).

Depletion of several nutrients may occur as a result of prolonged and heavy usage of antacids.

Hypophosphatemia has been reported with excessive use of antacids containing aluminum and magnesium hydroxide (Lotz et al. 1978). Symptoms of hypophosphatemia do not, however, usually occur unless the plasma phosphate is reduced below 0.35 mmol/l. Symptoms associated with phosphate depletion include anorexia, muscle weakness, tremors, paresthesiae, confusion, hemolytic anemia, abnormal leukocyte function, and coma (*Lancet* 1981). Secondary osteomalacia may develop, though this complication is uncommon. When it does occur, osteomalacia can cause bone pain and difficulty in walking (Insogna et al. 1980).

Patient groups who are at special risk for antacid-induced hypophosphatemia include the following:

1   Elderly patients who habitually take aluminum- or magnesium-containing antacids with food (Roe 1984)
2   Alcoholics prone to antacid abuse because of recurrent acute gastritis (Roe 1979)
3   Formula-fed patients who are receiving Al/Mg antacids with TPN or enteral feeding solutions low in phosphate (Grant and Todd 1982)
4   Renal patients who receive very high therapeutic doses of Al/Mg antacids in order to prevent hyperphosphatemia (Roe 1984)

Antacids can cause copper deficiency by precipitating dietary copper at an alkaline pH in the intestine. Signs of copper deficiency include

TABLE 6.1 COMMON DRUG GROUPS AND DRUGS THAT MAY CAUSE NUTRIENT DEPLETION AND NUTRITIONAL DEFICIENCIES

Drug Group	Drug	Deficiency
Antacids	Sodium bicarbonate Aluminum hydroxide	Folate, phosphate, calcium, copper
Anticonvulsants	Phenytoin, phenobarbital, primidone	Vitamins D and K
	Valproic acid	Carnitine
Antibiotics	Tetracycline	Calcium
	Gentamicin	Potassium, magnesium
	Neomycin	Fat, nitrogen
Antibacterial agents	Boric acid	Riboflavin
	Trimethoprim	Folate
	Isoniazid	Vitamin B_6, niacin, Vitamin D
Anti-inflammatory agents	Sulfasalazine	Folate
	Aspirin	Vitamin C, Folate, iron
	Colchicine	Fat, Vitamin B_{12}
	Prednisone	Calcium
Anticancer drugs	Methotrexate	Folate, calcium
	Cisplatin	Magnesium
Anticoagulants	Warfarin	Vitamin K
Antihypertensive agents	Hydralazine	Vitamin B_6
Antimalarials	Pyrimethamine	Folate
Diuretics	Thiazides	Potassium
	Furosemide	Potassium, calcium, magnesium
	Triamterene	Folate
H^2 receptor antagonists	Cimetidine Ranitidine	Vitamin B_{12}
Hypocholesterolemic agents	Cholestyramine	Fat
	Colestipol	Vitamin K, Vitamin A, folate, Vitamin B_{12}
Laxatives	Mineral oil	Carotene, retinol, Vitamins D, K
	Phenolphthalein	Potassium
	Senna	Fat, calcium
Oral contraceptives		Vitamin B_6, folate, Vitamin C
Tranquilizers	Chlorpromazine	Riboflavin

hypocupremia, anemia, leukopenia, altered iron metabolism, ataxia, osteoporosis, and connective tissue disorders.

The risk of copper deficiency increases when high and persistent levels of antacids are taken in conjunction with zinc supplements, which decrease copper absorption. Patients receiving antacids while on a copper-deficient TPN formula are at similar risk (*Nutr. Rev.* 1984).

Folate malabsorption can be induced by ingestion of antacids which render the pH of the jejunum more alkaline. In all studies in which this has been demonstrated, the antacid used was sodium bicarbonate (Benn et al. 1971; MacKenzie and Russell 1976).

Patients with pancreatic insufficiency due to cystic fibrosis or chronic alcoholic pancreatitis absorb folic acid better than normal control subjects (Russell et al. 1979). When patients with pancreatic insufficiency are treated with oral pancreatic extracts and sodium bicarbonate to manage their maldigestion, their ability to absorb folic acid is markedly reduced (Russell et al. 1980).

When pancreatic insufficiency is untreated, the pH of the jejunum is low and is favorable to absorption of folic acid. Treatment renders the pH of the jejunum alkaline and compromises folic acid uptake; formation of an insoluble complex between pancreatic extract and folate reduces absorption of the vitamin.

These findings indicate a need to monitor the folate status of patients with pancreatic insufficiency who are being treated with pancreatic extracts, and to give them folic acid supplements as necessary to prevent folate depletion and reduce the risk of their developing megaloblastic anemia.

Patients with peptic ulcer are not necessarily at risk for antacid-induced nutrient depletion because the preferred method of treatment of hyperchlorhydria is by administration of cimetidine or ranitidine, and because such patients may be given nutrient supplements containing folic acid (Bailey 1984).

ANTIBACTERIAL AGENTS

ANTIBIOTICS

Antibiotics may induce maldigestion and malabsorption of nutrients. They may produce vitamin antagonism or renal hyperexcretion of minerals secondary to nephrotoxicity.

Neomycin. Neomycin is a polybasic antibiotic used to reduce the growth of colonic bacteria prior to intestinal surgery, and in the treatment of impending or acute hepatic encephalopathy. At therapeutic doses, malabsorption of fats, nitrogen, glucose, beta-carotene, iron, and vitamin B_{12} occur.

Neomycin's effects on intestinal structure and gastroenterological function include shortening of villi, infiltration of the lamina propria with inflammatory cells, and crypt cell damage. Disaccharidase activity and micelle formation are inhibited (Holtzapple and Schwartz 1984; Jacobson et al. 1960; Cain et al. 1968; Thompson et al. 1970).

Aminoglycosides. The aminoglycoside antibiotics, such as gentamicin, are nephrotoxic. Gentamicin-induced renal tubular injury can lead to combined hypokalemia, hypomagnesemia, hypocalcemia, and alkalosis (Mazze and Cousins 1973).

Cephalosporins. Cephalosporins (beta-lactam antibiotics), such as moxalactam disodium, can produce a vitamin K–responsive coagulopathy. Spontaneous hemorrhage and bleeding into the gastrointestinal tract and epistaxis have been reported (Hooper et al. 1980; Pakter et al. 1982). These drugs can be considered to be vitamin K antagonists. They can also produce thrombocytopenia.

Tetracyclines. Tetracyclines may reduce iron absorption and, with prolonged usage, can contribute to calcium depletion (Neuvonen et al. 1970).

Other Antibacterial Agents. Trimethoprim, in combination with sulfamethoxazole in Bactrim (Roche), and Septra (Burroughs-Wellcome), is commonly used in the treatment of urinary tract infections and chronic bronchitis. Though it is a mild folate antagonist, trimethoprim presents minimal risk of folate deficiency (Girdwood et al. 1973).

ANTITUBERCULOSIS AGENTS

Isoniazid (isonicotinic acid hydrazide) has been used since the 1950s in the treatment of tuberculosis, and is the drug of choice for elderly TB patients. With chronic intake of this drug, patients are susceptible to peripheral neuropathy as a result of drug-induced vitamin B_6 deficiency (Nagami and Yoshikawa 1983). Side effects of the drug can be avoided by concurrent administration of vitamin B_6 at a dosage of 25 to 30 mg/day. It is not clear whether an observed susceptibility of elderly tuberculosis patients to isoniazid-induced vitamin B_6 depletion is related to slower metabolism of the drug in these patients (Yoshikawa and Nagami 1982).

Niacin deficiency can induce pellagra in patients receiving isoniazid who are also on a diet marginally deficient in niacin. Conversion of the amino acid tryptophan to niacin by the body is apparently inhibited by isoniazid (DiLorenzo 1967).

Isoniazid has also been shown to inhibit the hydroxylation of vitamin

D. Hepatic vitamin D 25-hydroxylase activity is reduced by isoniazid, and also by cimetidine (Bengoa et al. 1983).

There is a risk of metabolic bone disease associated with long-term use of isoniazid; the drug reduces synthesis of the active forms of vitamin D that are necessary for efficient calcium absorption (Brodie et al. 1981).

ANTICOAGULANTS

Coumarin anticoagulants are antivitamins that inhibit the conversion of precursor proteins into active clotting factors by vitamin K (Suttie 1973).

Vitamin K hydroquinone (vitamin KH_2) catalyzes carboxylation in the liver as vitamin K 2,3-epoxide. Vitamin K epoxide is recycled to vitamin K and vitamin KH_2 for reutilization. It has been shown that when coumarin or indandione anticoagulants are administered, vitamin K 2,3-epoxide accumulates in the liver. It has been proposed that these anticoagulants function by inhibiting regeneration of vitamin K and vitamin KH_2 from the epoxide (Willingham and Matschiner 1974; Bjornsson 1984).

A protein containing gamma-carboxyglutamic acid occurs in bones. When coumarin anticoagulants such as warfarin are administered during pregnancy warfarin embryopathy may result. This is characterized by hypoplastic nasal structure and multiple bone defects, including stippled calcification of the epiphyses of long bones (Warkany 1975).

It has been proposed that bone defects in warfarin embryopathy are due to decreased synthesis of osteocalcin, a vitamin K–dependent protein which plays a key role in the mineralization of the fetal skeleton (*Nutr. Rev.* 1979a). Prenatal exposure to phenytoin may cause vitamin K–dependent hemorrhagic disease of the neonate. When phenytoin is given, there is elevation of serum osteocalcin levels and though it is not established whether this reflects synthesis of partially carboxylated osteocalcin, it is thought that a change in osteocalcin levels is typical of the effects of phenytoin on a spectrum of vitamin K–dependent proteins (Keith et al. 1983).

ANTICONVULSANTS

Phenytoin, alone or in combination with phenobarbital and primidone, has been shown to cause folate depletion (Hoffbrand and Nicheles 1968). There is evidence that the drug can produce malabsorption of this B vitamin (Gerson et al. 1972), though the exact mechanism is still in doubt.

Phenytoin and phenobarbital have been shown to cause vitamin D deficiency with secondary calcium malabsorption. Reduced calcium absorption is related to long-term use of these drugs (Wahl et al. 1981). Development of rickets or osteomalacia, though associated with phenytoin use (Deut et al. 1970; Hahn et al. 1972), is related to the coexistence of other etiological factors such as lack of sunlight exposure which reduces cutaneous synthesis of vitamin D (Hahn and Avioli 1984).

Hemorrhage may occur in the newborn infants of women who have received these anticonvulsant agents during pregnancy. Such a coagulation defect is prevented by administration of vitamin K to the mother or to the neonate (Mountain et al. 1970). Water-soluble preparations of vitamin K are contraindicated because they can cause hemolytic anemia and transport of bile pigment into the brain (Kernicterus) of the neonate (Evans et al. 1970).

ANTIHYPERTENSIVE AGENTS

Hydralazine is a vitamin B_6 antagonist. Kirkendall and Page (1958), who first reported this adverse nutritional outcome of hydralazine use, emphasized that the neuropathy occurred in malnourished patients and was resolved by administration of pyridoxine and stopping the drug.

In another report (Raskin and Fishman, 1965), it was estimated that about 10% of patients receiving hydralazine show symptoms of vitamin B_6 deficiency. The antipyridoxic effect of the drug is due both to the formation of a pyridoxal–hydralazine complex and to drug-induced inhibition of pyridoxal kinase.

ANTI-INFLAMMATORY AGENTS

ASPIRIN

Aspirin and indomethacin can cause iron-deficiency anemia by inducing blood loss from the gastrointestinal tract. Erosions in the gastrointestinal mucosa and prolongation of bleeding time with high aspirin intake in certain individuals have been implicated (Leonards and Levy 1973; Quick 1966).

Aspirin in therapeutic doses can reduce serum folate. The drop in serum folate is rapidly reversed after aspirin intake is discontinued

(Lawrence et al. 1984). Chronic low serum folate values found in patients with rheumatoid arthritis may in part be explained by their use of aspirin (Alter et al. 1971).

Aspirin can also produce vitamin C (ascorbic acid) depletion (Coffey and Wilson 1975). Sahud and Cohen (1971) found that plasma ascorbic acid levels were abnormally low in patients with rheumatoid arthritis unless supplementary vitamin C was given. It has been demonstrated in vitro that aspirin blocks the uptake of ascorbic acid into blood platelets (Sahud 1970). Aspirin also increases the urinary excretion of ascorbic acid (Daniels and Everson 1937).

Aspirin may also affect intestinal function. For example, the drug's inhibitory effect on glucose absorption may in part explain lower blood glucose values found with aspirin use (Arvanitakis et al. 1977).

Colchicine. Colchicine, long used in the treatment of gout, frequently induces gastrointestinal side effects (Race et al. 1970). It was found that the drug causes malabsorption and produces increased fecal losses of fat and nitrogen. Malabsorption of vitamin B_{12} is also induced by colchicine. Because colchicine is usually employed for short-term therapy, the risk of inducing vitamin B_{12} deficiency with this drug is remote.

Sulfasalazine. Sulfasalazine is an anti-inflammatory agent used in the treatment of ulcerative colitis and regional enteritis. It inhibits the intestinal absorption of dietary folate because of competition for the folate transport system. In vitro, sulfasalazine inhibits dihydrofolate reductase, methyltetrahydrofolate reductase, and serine transhydroxymethylase, all of which are involved in dependent metabolism (Halsted et al. 1981; Baum et al. 1981).

In clinical studies it has been found that sulfasalazine impairs folate absorption, though folate deficiency is most likely to occur when other causes of folate depletion are also present: dietary deficiency, celiac disease, hemolytic anemia, and regional enteritis with involvement of the small intestine (Swinson et al. 1981; Kane and Boots 1977; Schneider and Beeley 1977).

Penicillamine. Penicillamine, used for many years as a chelating agent, has also been found effective as an anti-inflammatory agent in rheumatoid arthritis. The incidence of side effects of the drug is high, among them a number of adverse nutritional effects (Stein et al. 1980). The chelating effect of the drug may produce zinc deficiency, and may in part explain observed dysgeusia. However, since the drug also causes stomatitis, dysgeusia may also be related to mucosal changes. It has been suggested that taste impairment induced by the drug could be due to copper chelation (Lyle 1974; Henkin et al. 1967).

Penicillamine is also a vitamin B_6 antagonist. Effects of the drug on vitamin B_6 metabolism may include formation of a complex with pyridoxal phosphate and inhibition of pyridoxal kinase (Rumsby and Shepherd 1981).

Glucocorticoids. Steroidal anti-inflammatory agents decrease calcium absorption and increase calcium excretion. Demineralization and fractures may accompany drug-induced osteoporosis with prolonged use. A higher risk of corticosteroid-induced osteoporosis is defined in postmenopausal women, in alcoholics, and in the elderly (Roe 1985; 1981a).

Very high doses of corticosteroids can cause pancreatic necrosis, maldigestion, and malabsorption (Kimburg 1969).

ANESTHETICS

Nitrous oxide, previously used only as an anesthetic gas but now used in pulmonary ventilation after cardiac bypass surgery, interferes with the function of vitamin B_{12}, causing reversible megaloblastosis. Nitrous oxide oxidizes vitamin B_{12} in vitro and in vivo when premixed as a 50% mixture with oxygen as it is used in practice (Amess et al. 1978).

CANCER CHEMOTHERAPEUTIC AGENTS

Cancer chemotherapeutic agents can cause nutritional deficiencies from the following drug-related toxic effects:

1   Vitamin antagonism causing impaired vitamin utilization
2   Nephrotoxicity with renal depletion of minerals
3   Systemic toxicity with anorexia and vomiting, leading to reduced food intake and retention.

Vitamin Antagonism.   Acute folate deficiency occurs when high doses of the folate antagonists used in cancer and leukemia therapy are administered. Methotrexate, which is the folate antagonist or antifol used most often in the treatment of malignant disease, enters cells and binds firmly to the dihydrofolate reductase enzyme which is normally responsible for the metabolic interconversion of folate. When the drug becomes bound, folate is displaced from the enzyme and excreted in the urine. Further, methotrexate polyglutamates are formed and there is reduced synthesis of polyglutamates derived from dietary folacin. Thymidylate synthetase is inhibited and there is then inhibition of DNA, RNA, and protein synthesis. Cytotoxic effects of the drug are selective according to the rate of cell reproduction (Hoffbrand 1975).

The cytotoxic effects of methotrexate are particularly manifested on the gastrointestinal tract. Mucosal ulceration occurs following short-term, high-dose administration of the drug; with long-term use, there may be malabsorption with steatorrhea and secondary calcium depletion (Roe 1983). Methotrexate is an example of the class of antifols that have a close structural resemblance to folate and that are in the group of aminofols.

Another class of antifols are the diaminopyrimidines, of which the best known is the antimalarial agent pyrimethamine. Triazinate, an experimental cancer chemotherapeutic agent, is another drug in this category. Triazinate is administered by the intravenous route because it is poorly absorbed from the gastrointestinal tract. Signs of acute folate deficiency develop when triazinate is given in therapeutic doses (Skeel et al. 1976). In addition to its gastrointestinal side effects that are identical to those produced by methotrexate, cutaneous side effects are common. These include exfoliative dermatitis and also marked flexural hyperpigmentation, resembling acanthosis nigricans. It has been suggested that the hyperpigmentation produced by this drug is associated with folate deficiency (Greenspan et al. 1985).

The cardiotoxicity of adriamycin has been attributed to drug-induced vitamin E deficiency. However, although in experimental animals the cardiotoxicity of adriamycin has been ameliorated by administration of vitamin E, vitamin E has not prevented the toxicity of the drug in cancer patients (Myers et al. 1976; Pratt and Rudden 1979).

Nephrotoxic Effects.   The cancer chemotherapeutic agent Cisplatin (Cis-Platinum [II] dichlorodiamine) is nephrotoxic and can cause magnesium depletion. The nephrotoxicity of cis-platinum is dose dependent and it has been suggested that its nephrotoxic effects may be influenced by the concurrent administration of other cancer chemotherapeutic drugs (Buckley et al. 1984). The nephrotoxic effects of these drugs have been reduced by adequate hydration of patients prior to therapy (Schilsky and Anderson 1979; Stark and Howell 1978).

Systemic Toxicity.   The adverse nutritional effects of cancer chemotherapeutic agents vary in impact according to the pretherapy nutritional status of the patient and the type of nutritional support given during therapy. Anorexia, nausea, and vomiting are more severe if the regime is intensive and prolonged (Shils 1979) and are greatly dependent on dosage schedule.

DIURETICS

Diuretics can cause loss of minerals, in particular sodium, to the extent of hyponatremia (Thomas et al. 1978). Oral administration of diuretics such as furosemide, ethacrynic acid, and triamterene cause significant hypercalciuria. Renal clearance of calcium is increased by intravenous or oral furosemide, intravenous ethacrynic acid, or oral triamterene. Thiazide drugs, on the other hand, may cause a mild hypercalcemia (Eknoyan et al. 1970; Duarte et al. 1971).

Potassium depletion may occur with prolonged or high dosages of oral diuretics, including the thiazides and the loop diuretics, furosemide and ethacrynic acid (Steen 1981; Hamdy et al. 1980). Triamterene and the aldosterone antagonist, spironolactone, are potassium-sparing diuretics (Facts and Comparisons 1984).

Diuretics may introduce susceptibility to digitalis intoxication with arrythmia, development of nondigitalis-related arrythmias, impairment in renal tubular function, hyperglycemia, and muscle weakness characteristic of potassium deficiency (Steiness 1981).

H_2-RECEPTOR ANTAGONISTS

The H_2-receptor antagonists cimetidine and ranitidine cause reversible malabsorption of protein-bound vitamin B_{12} (cobalamin). Absorption of unbound cobalamin is maintained within normal limits.

The effects of these drugs on pepsin and gastric acid secretion are such that the vitamin B_{12} in animal protein food is not freed from its protein binder during digestion (Steinberg et al. 1980; Streeter et al. 1982; Lelaiche 1983).

Cimetidine, but not ranitidine, enhances the hypoprothrombinemic effect of warfarin and other coumarin anticoagulants. These effects are due to inhibition of anticoagulant metabolism by the drug (Serlin et al. 1979; 1981).

HYPOCHOLESTEROLEMIC AGENTS

Cholestyramine is a bile acid sequestrant that causes malabsorption when given in doses greater than 16 g/day. Malabsorption of fat-soluble vitamins may develop with prolonged usage. The drug can lead to folate depletion and imposes a risk for the development of megaloblastic anemia if intake of folate is low. Colestipol can cause malabsorption of fat-soluble vitamins by the same mechanism (Glueck et al. 1972; West and Lloyd 1975).

LAXATIVES

Laxatives can be classified into four groups according to a modification of the system proposed by Fingl and Preston (1979). These four groups, based on the primary mode of action, are as follows:

Agents that Alter Electrolyte Transport. Anthraquinone drugs such as senna and cascara, and phenylmethane derivatives such as phenolphthalein, reduce sodium absorption and convert sodium and water absorption into net secretion in the large intestine. Evidence suggests that they also affect intestinal mucosa permeability and increase intestinal motility.

Ricinoleic acid (the active ingredient of castor oil) and dioctylsulfosuccinate are surface-active agents that also inhibit the absorption of sodium and convert sodium and water absorption into net secretion, but their mechanism of action is different. They stimulate adenylate cyclase, inhibit sodium potassium ATPase, and increase membrane permeability. They also increase the hydration of the feces (Hardcastle and Wilkins 1970).

Saline Cathartics. Saline cathartics include magnesium and sodium sulfates. They are primarily osmotic laxatives that function by drawing water into the colon. They may also trigger release of cholecystokinin into the duodenum. Secretion in the pancreas and the small and large intestines may increase (Ganginella and Phillips 1975; Donowitz and Binder 1975; Harvey and Read 1975).

Stool Softeners. Mineral oil is a mixture of liquid hydrocarbons obtained from petroleum. The oil is indigestible and is not absorbed from the intestine. It functions primarily as a stool softener but also may inhibit water reabsorption from the colon (Fingl 1975).

Bulking Agents. Natural and synthetic polysaccharides are used as bulking agents. Those in common usage include bran, methylcellulose, ispaghula, and psyllium seed. Bulk-forming agents not only increase the size of the stool, but also decrease the transit time of the contents of the large intestine and delay stomach emptying. Although this action does not contribute to their laxative effect, it does explain certain influences of these agents on nutrient absorption (Brodribb 1977).

Laxatives in all these groups can adversely affect nutrient absorption or can increase nutrient losses, particularly mineral losses, if taken in excess. Senna, phenolphthalein, and bisacodyl (in the first group) have all been found to induce potassium depletion and can cause malabsorption when taken by laxative abusers (Fleming et al. 1975; Levine et al. 1981).

Osteomalacia has been reported as a rare but serious adverse outcome of malabsorption resulting from laxative abuse (Frame et al. 1971). Abuse of laxatives can cause protein deficiency as a consequence of a protein-losing enteropathy (Heizer et al. 1968).

Laxative abuse is common in three patient groups: girls and women with anorexia nervosa, depressed patients preoccupied with bowel function, and constipated elderly patients (Roe 1984; 1985). Massive diarrhea associated with castor oil or saline cathartics may cause hypokalemia but not malabsorption or protein-losing enteropathy (Race et al. 1970).

Mineral oil, alone or formulated as an emulsion with the addition of other laxatives such as phenolphthalein, can cause malabsorption of the fat-soluble vitamins, particularly beta-carotene and vitamins A, D, and K (Morgan 1941; Curtis and Balmer 1939).

Bulking agents in general do not have an adverse effect on nutritional status. However, ingestion of very large amounts of psyllium seed with food can reduce riboflavin absorption (Roe, unpublished). No systematic studies of the effects of bulking agents on nutrient absorption have been carried out in human subjects.

ORAL CONTRACEPTIVES

Many adverse nutritional effects of oral contraceptives were identified at the time when high-estrogen preparations were in use (Roe 1977). Vitamin status changes have not been found to occur with the low-estrogen preparations now in general use (Roe et al. 1982).

Folate.   When folate-responsive megaloblastic anemia has occurred in women receiving oral contraceptives, other etiological factors have also been present including dietary folate deficiency and malabsorption (Streiff 1970; Stephens et al. 1972; Toghill and Smith 1971; Wood et al. 1972; Holtzapple and Schwartz 1984).

In a metabolic study comparing the folate status of oral contraceptive users and nonusers, the folacin intake of both groups was held constant throughout the investigation. There were individual differences in plasma and red cell folate values after subjects had been on the diet eight weeks. Plasma and red cell folacin were slightly lower in the oral contraceptive users. When the oral contraceptive users were divided into those who had been on these drugs for less than six months and those who had been on the drugs longer than six months, the longer-term OC users had lower plasma and red cell folate values but none of these differences was found to be statistically significant (Roe 1981a).

In a study of racial differences among adolescents taking oral contraceptives, Grace et al. (1982) found a significant decrease in mean serum folate and whole blood folate values in the White but not in the Black group. This difference was not explained by diet and indeed the investigators were unable to explain the findings.

Whether there is a need to give folic acid supplements to oral contraceptive users has been a subject of much discussion. On the basis of studies of pregnant women in Sri Lanka, Hettiarachchy et al. (1983) concluded that recent oral contraceptive users need folic acid supplements during pregnancy to offset folate depletion occurring during gestation.

Megaloblastic changes have been reported in smears of cervical epithelial cells obtained from women on oral contraceptives; these changes disappeared following administration of folic acid supplements (Lindenbaum et al. 1975). Subsequently, Butterworth et al. (1982) showed that the cervical megaloblastosis occurring with oral contraceptive use is associated with cervical dysplasia, and that both conditions were resolved when folic acid supplements were given.

Vitamin B$_{12}$. Shojania and Wylie (1979) examined the effects of oral contraceptives on vitamin B$_{12}$ status. Nineteen of 199 women receiving combination or sequential oral contraceptives had serum levels of vitamin B$_{12}$ significantly below the normal range. The fall in serum vitamin B$_{12}$ was attributed to changes in the vitamin's binders in the serum.

Other B Vitamins. Effects of oral contraceptive drugs on thiamin, riboflavin, and pantothenic acid status have been studied in metabolic units with strict control of dietary intake. Under these conditions Lewis and King (1980) found no significant effect.

When we examined the riboflavin requirements of oral contraceptive users and nonusers, there were no significant intergroup differences attributable to the Pill (Roe et al. 1982).

However, these findings stand in contrast to those by Donald and Bosse (1979). These investigators found that oral contraceptive users may need at least 1.5 mg of pyridoxine per day to achieve normal vitamin B$_6$ status. Depression associated with oral contraceptives may be alleviated by vitamin B$_6$, according to a review of the subject (*Nutr. Rev.* 1979b).

Vitamin E. Oral contraceptives were not shown to cause vitamin E depletion. Marginally inadequate vitamin E status was found in OC users and nonusers of the group studied (Tongney and Driskell 1978).

Vitamin D. Effects of oral contraceptives on vitamin D status have also been examined and no significant effect was identified (Schreurs et al. 1981).

REFERENCES

ALTER, H.J., N.J. ZVAIFLER, and C.E. RATH. 1971. Interrelationships of rheumatoid arthritis, folic acid and aspirin. *Blood* 38: 405–16.

AMESS, J.A., J.F. BURMAN, G.M. REES, D.G. NANCEKIEVILL, and D. MOLLIN. 1978. Megaloblastic haemopoiesis in patients receiving nitrous oxide. *Lancet* 2: 339–42.

ARVANITAKIS, C., G-H. CHEN, J. FOLSCROFT, and N.J. GREENBERGER. 1977. Effect of aspirin on intestinal absorption of glucose, sodium and water in man. *Gut* 18: 187–90.

BAILEY, A. 1984. The Effects of Antacid Consumption on Folate Status. M.S. Thesis, Cornell University, Ithaca, NY.

BAUM, C.L., J. SELHUB, and I.H. ROSENBERG. 1981. Antifolate actions of sulfasalazine on intake of lymphocytes. *J. Lab. Clin. Med.* 97: 778.

BENGOA, J.M., M.J.G. BOLT, and I.H. ROSENBERG. 1983. Hepatic vitamin D 25-hydroxylase inhibition by cimetidine and isoniazid. *Gastroenterology* 84:1363.

BENN, A., C.J.H. SWAN, W.T. COOKE, J.A. BLAIR, A.J. MATTY, and M.E. SMITH. 1971. Effect of intraluminal pH on the absorption of pteroylmonoglutamic acid. *Brit. Med. J.* 16:148–50.

BJORNSSON, T.D. 1984. Vitamin K and vitamin K antagonists. In *Drugs and Nutrients: The Interactive Effects.* ed. D.A. Roe and T.C. Campbell. 429–73. New York: Marcel Dekker.

BRODIE, M.J., A.R. BOOBIS, C.J. HILLYARD, G. ABEYASEKERA, I. MACINTYRE, and B. KEVIN PARK. 1981. Effect of isoniazid on vitamin D metabolism and hepatic homooxygenase activity. *Clin. Pharmacol. Ther.* 30: 363–67.

BRODRIBB, A.J.M. 1977. Treatment of symptomatic diverticular disease with a high fiber diet. *Lancet* 1: 664–66.

BUCKLEY, J.E., V.L. CLARK, T.J. MEYER, and N.W. PEARLMAN. 1984. Hypomagnesemia after cisplatin combination chemotherapy. *Ann. Intern. Med.* 144: 2347–48.

BUTTERWORTH, C.E., K.D. HATCH, H. GORE, H. MUELLER, and C.L. KRUMDIECK. 1982. Improvement in cervical dysplasia associated with folic acid therapy in users of oral contraceptives. *Am. J. Clin. Nutr.* 35: 73–82.

CAIN, G.D., E.B. REINER, and M. PATTERSON. 1968. Effects of neomycin on disaccharidase activity of the small bowel. *Arch. Intern. Med.* 122: 314.

COFFEY, G. and C.W.M. WILSON. 1975. Ascorbic acid deficiency and aspirin-induced haematemesis. *Brit. Med. J.* 1: 108.

CURTIS, A.C. and R.S. BALMER. 1939. The prevention of carotene absorption. *J.A.M.A.* 113: 1785–88.

DANIELS, A.L. and G.J. EVERSON. 1937. Influence of acetylsalicylic acid (aspirin) on urinary excretion of ascorbic acid. *Proc. Soc. Exp. Biol. Med.* 35: 20–24.

DENT, C.E., A. RICHENS, D.J.F. ROWE, and T.C.B. STAMP. 1970. Osteomalacia with long term anticonvulsant therapy in epilepsy. *Brit. Med. J.* 4: 69–72.

DILORENZO, P.A. 1967. Pellagra-like syndrome associated with isoniazid therapy. *Acta Dermatol. Venereol.* 47: 318–22.

DONALD, E.A. and T.R. BOSSE. 1979. The vitamin B_6 requirement in oral contraceptive users. I. Assessment by pyridoxal level and transferase activity in erythrocytes. *Am. J. Clin. Nutr.* 32.

DONOWITZ, M. and H.J. BINDER. 1975. Effect of dioctyl sulfosuccinate on colonic fluid and electrolyte movement. *Gastroenterology* 69: 941–50.

DUARTE, C.G., J.L. WINNAKER, K.L. BECKER, and A. PAGE. 1971. Thiazide induced hypercalcemia. *New Engl. J. Med.* 284: 828.

EKNOYAN, G., W.N. SUKI, and M. MARTINEZ-MALDONARDO. 1970. Effect of diuretics on urinary excretion of phosphate, calcium, and magnesium in thyroparathyroidectomized dogs. *J. Lab. Clin. Med.* 76: 25.

EVANS, A.R., R.M. FORESTER, and C. DISCOMBE. 1970. Neonatal hemorrhage following maternal anticonvulsant therapy. *Lancet* 1: 517–18.

FACTS AND COMPARISONS. 1984. St. Louis: J.B. Lippincott.

FINGL, E. 1975. Laxatives and cathartics. In *The Pharmacological Basis of Therapeutics*, 5th ed. ed. L.S. Goodman and A. Gilman. 978. New York: Macmillan.

FINGL, E. and J.W. PRESTON. 1979. Antidiarrheal agents and laxatives: changing concepts. *Clin. Gastroenterol.* 8: 161–85.

FLEMING, B.J., S.M. GENUTH, A.B. GOULD, and M.D. KAMINOKOWSKI. 1975. Laxative induced hypokalemia, sodium depletion, and hyperrreninemia. Effects of potassium and sodium replacement on the reninangiotensin system. *Ann. Intern. Med.* 83: 60–62.

FRAME, B., H.L. GUIANG, H.M. FROST, and W.A. REYNOLDS. 1971. Osteomalacia induced by laxative (phenolphthalein) ingestion. *Arch. Intern. Med.* 128: 794–96.

GANGINELLA, T.S. and S.F. PHILLIPS. 1975. Ricinoleic acid: Current view of an ancient oil. *Am. J. Dis. Dig.* 20: 1171–77.

GERSON, C.D., G.W. HEPNER, N. BROWN, H. COHEN, V. HERBERT, and H.D. JANOWITZ. 1972. Inhibition by diphenylhydantoin of folic acid absorption. *Gastroenterology* 65: 246–51.

GIRDWOOD, R.H., A.J. DACOSTA, and R.R. SAMSON. 1973. Co-trimoxazole as a possible cause of folate depletion. *Brit. J. Haematol.* 25: 279–80.

GLUECK, C.J., S. FORD, D. SCHEEL, and P. STEINER. 1972. Colestipol and cholestyramine resin. Comparative effects in familial type II hyperlipoproteinemia. *J.A.M.A.* 222: 676.

GRACE, E., S.J. EMANS, and D.E. DRUM. 1982. Hematological abnormalities in adolescents who take oral contraceptives. *J. Pediatr.* 101: 771–74.

GRANT, A. and E. TODD. 1982. *Enteral and Parenteral Nutrition, A Clinical Handbook.* p. 148. Oxford and London: Blackwell.

GREENSPAN, A.H., J.L. SHUPACK, S-H. FOO, and A.C. WISE. 1985. Acanthosis nigricans–like hyperpigmentation secondary to triazinate therapy. *Arch. Dermatol.* 121: 232–35.

HAHN, T.J. and L.V. AVIOLI. 1984. Anticonvulsant-drug-induced mineral disorders. In *Drugs and Nutrients: The Interactive Effects.* ed. D.A. Roe and T.C. Campbell. 409–27. New York: Marcel Dekker.

HAHN, T.J., S.J. BIRGE, C.R. SCHARP, and L.V. AVIOLI. 1972. Phenobarbital-induced alterations in vitamin D metabolism. *J. Clin. Invest.* 51: 741–48.

HALSTED, C.H., G. GANDHI, and T. TAMURA. 1981. Sulfasalazine inhibits the absorption of folates in ulcerative colitis. *New Engl. J. Med.* 305: 1513–16.

HAMDY, R.C., J. TOVEY, and N. PERERA. 1980. Hypokalemia and diuretics. *Brit. Med. J.* 1: 1187.

HARDCASTLE, J.D. and J.L. WILKINS. 1970. The action of sennosides and related compounds on human colon and rectum. *Gut* 11: 1038–42.

HARVEY, R.F. and A.E. READ. 1975. Mode of action of the saline purgatives. *Am. Heart J.* 89: 810–12.

HEIZER, W.D., A.L. WARSHAW, T.A. WALDMANN, and L. LASTER. 1968. Protein-losing gastroenteropathy and malabsorption associated with factitious diarrhea. *Arch. Intern. Med.* 68: 839–51.

HENKIN, R.I., H.R. KEISER, I.A. JAFFE, I. STERNLIEB, and I.H. SCHEINBERG. 1967. Decreased taste sensitivity after D-penicillamine reversed by copper administration. *Lancet* 2: 1268.

HETTIARACHCHY, N.S., S.S. SRIKANTHA, and S.M.X. COREA. 1983. The effect of oral contraceptive therapy and pregnancy on serum folate levels of rural Sri Lankan women. *Brit. J. Nutr.* 50: 495–501.

HOFFBRAND, A.V. 1975. Synthesis and breakdown of natural folates (folate polyglutamates). In *Progress in Hematology,* vol. 9, ed. E.B. Brown. 85–105. New York: Grune and Stratton.

HOFFBRAND, A.V. and R.F. NICHELES. 1968. Mechanisms of folate deficiency in patients receiving phenytoin. *Lancet* 2: 528–30.

HOLTZAPPLE, P.G. and S.E. SCHWARTZ. 1984. Drug-induced maldigestion and malabsorption. In *Drugs and Nutrients: The Interactive Effects.* ed. D.A. Roe and T.C. Campbell. 475–85. New York: Marcel Dekker.

HOOPER, C.A., B.B. HANEY, and H.H. STONE. 1980. Gastrointestinal bleeding due to vitamin K deficiency in patients on parenteral cefamandole. *Lancet* 1: 39–40.

INSOGNA, K.L., D.R. BORDLY, J.F. CARO, and D.H. LOCKWOOD. 1980. Osteomalacia and weakness from excessive antacid. *J.A.M.A.* 244: 2544–46.

JACOBSON, E.D., J.T. PRIOR, and W.W. FALOON. 1960. Malabsorptive syndrome induced by neomycin: morphologic alterations in the jejunal mucosa. *J. Lab. Clin. Med.* 56: 245–50.

KANE, S.P. and M.A. BOOTS, 1977. Megaloblastic anaemia associated with sulphasalazine treatment. *Brit. Med. J.* 2: 1287–88.

KEITH, D.A., C.M. GUNDBERG, A. JAPOUR, J. ARONOFF, N. ALVAREZ, and GALLOP. 1983. *Clin. Pharmacol. Ther.* 34: 529–32.

KIMBURG, D.V. 1969. Effect of vitamin D and steroid hormones on the active transport of calcium by the intestine. *New Engl. J. Med.* 280: 1396–1405.

KIRKENDALL, W.M. and E.B. PAGE. 1958. Polyneuritis occurring during hydralazine therapy. *J.A.M.A.* 167: 427–32.

LANCET. 1981. Editorial. Treatment of severe hypophosphataemia. *Lancet* 2: 734.

LAWRENCE, V.A., J.E. LOEWENSTEIN, and E.R. EICHNER. 1984. Aspirin and folate binding: in vivo and in vitro studies of serum binding and urinary excretion of endogenous folate. *J. Lab. Clin. Med.* 103: 944–48.

LELAICHE, J., D. COTTON, J. ZITTONN, J. MARQUET, and I. YVART. 1983. Effect of ranitidine on cobalamin absorption. *Dig. Dis. Sci.* 28: 667.

LEONARDS, J.H. and G. LEVY. 1973. Gastrointestinal blood loss during prolonged aspirin administration. *New Engl. J. Med.* 289: 1020–22.

LEVINE, D., E.A.W. GOOD, and D.L. WINGATE. 1981. Purgative abuse associated with reversible cachexia, hypogammaglobulinemia and finger clubbing. *Lancet* 1: 919–20.

LEWIS, C.M. and J.C. KING. 1980. Effect of oral contraceptive agents on thiamin, riboflavin and pantothenic acid status in young women. *Am.'J. Clin. Nutr.* 33: 832–38.

LINDENBAUM, J., N. WHITEHEAD, and F. REYNER. 1975. Oral contraceptive hormones, folate metabolism, and the cervical epithelium. *Am. J. Clin. Nutr.* 28: 346–53.

LOTZ, M., E. ZISMAN, and C. BARTTER. 1978. Evidence for a phosphorus depletion syndrome in man. *New Eng. J. Med.* 278: 409–15.

LYLE, W.H. 1974. Penicillamine and zinc. *Lancet* 2: 1140.

MACKENZIE, J.F. and R.J. RUSSELL. 1976. The effect of pH on folic acid absorption. *Clin. Sci. Molec. Med.* 51: 363–68.

MAZZE, R.I. and M.J. COUSINS. 1973. Combined nephrotoxicity of gentamicin and methoxyflurane anesthesia. A case report. *Brit. J. Anesthesiol.* 45: 394.

MORGAN, J.W. 1941. The harmful effects of mineral oil (liquid petrolatum) purgatives. *J.A.M.A.* 117: 1335–36.

MOUNTAIN, K.R., J. HIRSH, and A.S. GALLUS. 1970. Neonatal coagulation defect due to anticonvulsant drug treatment in pregnancy. *Lancet* 1: 265–68.

MYERS, C.E., W.M. McGUIRE, and R. YOUNG. 1976. Adriamycin: amelioration of toxicity by alpha-tocopherol. *Cancer Treat. Rep.* 60: 961.

NAGAMI, P.H. and T.T. YOSHIKAWA. 1983. Tuberculosis in the geriatric patient. *J. Am. Geriat. Soc.* 31: 356–63.

NEUVONEN, T.J., G. GOTHANI, and R. HACKMAN. 1970. Interference of iron with the absorption of tetracyclines in man. *Brit. Med. J.* 4: 532–34.

NUTRITION REVIEWS. 1979a. Editorial. Osteocalcin: A vitamin K–dependent calcium-binding protein in bone. *Nutr. Rev.* 37: 55–57.

NUTRITION REVIEWS. 1979b. Editorial. The vitamin B_6 requirement in oral contraceptive users. *Nutr. Rev.* 37: 344–45.

NUTRITION REVIEWS. 1984. Editorial. Conditioned copper deficiency due to antacids. *Nutr. Rev.* 42: 319–21.

PAKTER, R.L., T.R. RUSSELL, H. MIELKE, and D. WEST. 1982. Coagulopathy associated with the use of moxalactam. *J.A.M.A.* 248: 1100.

Physicians' Desk Reference for Non-Prescription Drugs. Oradell, NJ: Cole Economics.

PRATT, W.B. and R.W. RUDDEN. 1979. *The Anticancer Drugs.* 164–70. Oxford: Oxford Univ. Press.

QUICK, A.J. 1966. Salicylates and bleeding: the aspirin tolerance test. *Am. J. Med. Sci.* 252: 265.

RACE, T.F., I.C. PAES, and W.W. FALOON. 1970. Intestinal malabsorption induced by oral colchicine. Comparison with neomycin, and cathartic agents. *Am. J. Med. Sci.* 259: 32–41.

RASKIN, N.H. and R.A. FISHMAN. 1965. Pyridoxine-deficiency neuropathy due to hydralazine. *New Engl. J. Med.* 273: 1182–85.

ROE, D.A. 1977. Nutrition and the contraceptive pill. In *Nutritional Disorders of American Women.* ed. M. Winick. 37–49. New York: Wiley.

_____. 1979. *Alcohol and the Diet.* Westport, CT: AVI.

_____. 1981a. Intergroup and intragroup variables affecting interpretation of studies of drug effects on nutritional status. In *Nutrition in Health and Disease and International Development.* Symp. 12th. Internat. Congr. Nutrition. 757–71. New York: Alan R. Liss.

_____. 1981b. Drug interference with the assessment of nutritional status. *Clin. Lab. Med.* 1: 647–64.

_____. 1983. Drugs and nutrient absorption. In *Nutrition and Drugs.* ed. M. Winick. 129–38. New York: Wiley.

_____. 1984. Adverse nutritional effects of OTC drug use in the elderly. In *Drugs and Nutrition in the Geriatric Patient.* ed. D.A. Roe. New York: Churchill Livingstone.

_____. 1985. *Drug-Induced Nutritional Deficiencies,* 2nd ed. Westport, CT: AVI.

ROE, D.A., S. BOGUSZ, J. SHEU, and D.B. McCORMICK. 1982. Factors affecting riboflavin requirements of oral contraceptive users and non-users. *Am. J. Clin. Nutr.* 35: 495–501.

RUMSBY, P.C. and D.M. SHEPHERD. 1981. The effect of penicillamine on vitamin B_6 function in man. *Biochem. Pharmacol.* 30: 3051–53.

RUSSELL, R.M., G.J. DHAR, S.K. DUTTA, and I.H. ROSENBERG. 1979. Influence of intraluminal pH on folate absorption: studies in control subjects and in patients with pancreatic insufficiency. *J. Lab. Clin. Med.* 93: 428–36.

RUSSELL, R.M., S.K. DUTTA, E.V. OAKS, I.H. ROSENBERG, and A.G. GIOVETTI. 1980. Impairment of folic acid absorption by oral pancreatic extracts. *Dig. Dis. Sci.* 25: 369–73.

SAHUD, M.A. 1970. Uptake and reduction of dehydroascorbic acid in human platelets. Abstract. *Clin. Res.* 18: 133.

SAHUD, M.A. and R.J. COHEN. 1971. Effect of aspirin ingestion on ascorbic acid levels in rheumatoid arthritis. *Lancet* 1: 937–38.

SCHILSKY, R. and T. ANDERSON. 1979. Hypomagnesemia, secondary to cisdiammine-chloroplatinum II administration. *Ann. Intern. Med.* 90: 929–31.

SCHNEIDER, R.E. and L. BEELEY, 1977. Megaloblastic anaemia associated with sul-fasalazine treatment. *Brit. Med. J.* 1: 163–69.

SCHREURS, W.H.P., H.J.M. VAN RIJN, and H. VAN DEN BERG. 1981. Serum 25-hydroxycholecalciferol levels in women using oral contraceptives. *Contraception* 23: 399–406.

SERLIN, M.J., R.G. SIBEON, and A.M. BRECKENRIDGE. 1981. Lack of effect of ranitidine on warfarin action. *Brit. J. Clin. Pharmacol.* 12: 791.

SERLIN, M.J., R.G. SIBEON, S. MOSSMAN, A.M. BRECKENRIDGE, J.R.B. WILLIAMS, J.L. ATWOOD, and J.M.T. WILLOUGHBY, 1979. Cimetidine: Interaction with oral anticoagulants in man. *Lancet* 2: 317.

SHILS, M.E. 1979. Nutritional problems induced by cancer. *Med. Clin. N. Amer.* 63: 1009–25.

SHOJANIA, A.M. and B. WYLIE. 1979. Effect of oral contraceptives on vitamin B_{12} metabolism. *Am. J. Obstet. Gynecol.* 135: 129–34.

SKEEL, R.T., A.R. CASHMORE, W.L. SAWICKI, and J.R. BERTINO. 1976. Clinical and pharmacological evaluation of triazinate in humans. *Cancer Res.* 36: 48–54.

STARK, J.J. and S.B. HOWELL. 1978. Nephrotoxicity of cisplatinum (II) dichlorodiammine. *Clin. Pharmacol. Ther.* 23: 461–66.

STEEN, B. 1981. Hypokalemia, clinical spectrum, and etiology. *Acta Med. Scand. Suppl.* 647: 61–66.

STEIN, H.B., A.C. PATTERSON, R.C. OFFER, J. ATKINS, A. TENFEL, and H.S. ROBINSON. 1980. Adverse effects of D-penicillamine in rheumatoid arthritis. *Ann. Intern. Med.* 92: 24–29.

STEINBERG, W.M., C.E. KING, and P.P. TOSKES. 1980. Malabsorption of protein bound cobalamin but not unbound cobalamin during cimetidine administration. *Dig. Dis. Sci.* 25: 188–91.

STEINESS, E. 1981. Diuretics, digitalis, and arrhythmias. *Acta Med. Scand. Suppl.* 647: 75–85.

STEPHENS, M.E.M., I. CRAFT, and J. PETERS. 1972. Oral contraceptives and folate metabolism. *Clin. Sci.* 42: 405–14.

STREETER, A.M., K.J. GOULSTON, F.A. BATHUR, R.S. HILMER, G.G. GRANE, and M.T. PHEILS. 1982. Cimetidine and malabsorption of cobalamin. *Dig. Dis. Sci.* 27: 13–15.

STREIFF, R. 1970. Folate deficiency and oral contraceptives. *J.A.M.A.* 214: 105–8.

SUTTIE, J.W. 1973. Vitamin K and prothrombin synthesis. *Nutr. Rev.* 31: 105–9.

SWINSON, C.M., J. PERRY, M. LUMB, and A.J. LEVI. 1981. Role of sulphasalazine in the aetiology of folate deficiency in ulcerative colitis. *Gut* 22: 456–61.

THOMAS, T.H., D.B. MORGAN, and R. SWAMINATHAN. 1978. Severe hyponatremia: a study of 17 patients. *Lancet* 1: 621–24.

THOMPSON, G.R., M. MACMAHON, and P. CLAES. 1970. Precipitation by neomycin compounds of fatty acid and cholesterol from mixed micellar solutions. *Eur. J. Invest.* 1: 40–47.

TOGHILL, F.J. and P.G. SMITH. 1971. Folate deficiency and the Pill. *Brit. Med. J.* 1: 608–9.

TONGNEY, C.C. and J.A. DRISKELL. 1978. Vitamin E status of young women on combined type oral contraceptives. *Contraception* 17: 499–512.

WAHL, T.O., A.H. GOBRITY, and B.P. LUKERT. 1981. Long-term anticonvulsant therapy and intestinal calcium absorption. *Clin. Pharmacol. Ther.* 30: 506–12.

WARKANY, J. 1975. A warfarin embryopathy? *Am. J. Dis. Child.* 129: 287–88.

WEST, R.J. and J.K. LLOYD. 1975. The effect of cholestyramine on intestinal absorption. *Gut* 16: 93.

WILLINGHAM, A.K. and J.T. MATSCHINER. 1974. Changes in phylloquinone epoxidase activity related to prothrombin synthesis and microsomal clotting activity in the rat. *Biochem. J.* 140: 435–41.

WOOD, J.K., A.H. GOLDSTONE, and N.C. ALLEN. 1972. Folic acid and the Pill. *Scand. J. Haematol.* 9: 539–44.

YOSHIKAWA, T.T. and P.H. NAGAMI. 1982. Adverse drug reactions in tuberculosis therapy. Risks and recommendations. *Geriatrics* 37: 61–68.

Chapter 7

NUTRIENTS AS SUPPLEMENTS, DRUGS, AND NOSTRUMS

In industrialized countries, vitamins and minerals have long been used as drugs as well as to meet nutritional requirements. Persuaded by health professionals, commercial advertising, faddist organizations, or friends, the public will take nutritional supplements to improve their overall destiny, or simply as a means of general health promotion. Dangers of misuse are being increasingly reported. Rational use of vitamins and minerals, over and above the amount consumed in the diet, may be needed to meet nutritional requirements, to treat vitamin-dependency syndromes, as emergency treatment of drug intoxication, or in the prevention or treatment of diseases.

The pervasive use of food supplements in the United States has raised comment from overseas visitors and social critics for many years. It is a curious phenomenon in this land of food plenty! Perhaps it reflects a desire of many people to influence destiny. Until now, we have been content to observe the phenomenon but, as educators and health providers, we must become familiar with the adverse effects reported about food supplements so that we can better guide our clients and patients in their proper use (Lamy 1982).

ESTIMATES OF USAGE

A recent review cites estimates indicating that up to 50% of adults in the United States take vitamin supplements regularly (Dubick and Rucker 1983).

A 1974–75 survey of prescription and over-the-counter (OTC) drug use in a Southern rural community revealed that 20% of 152 black respondents and 18.3% of 216 white respondents took vitamins at least once during four one-week periods (Gagnon et al. 1978).

A study of healthy elderly persons in Albuquerque, New Mexico,

105

showed that 57% of the men and 61% of the women were taking one or more vitamin or mineral supplements; 31% of these individuals were ingesting a daily multivitamin preparation and 95% of those individuals were taking one or more additional vitamin or mineral supplement (Garry et al. 1982). Although 90% of this sample of 270 healthy men and women over 60 years of age got their RDA for ascorbic acid from diet alone, the median supplemental intake of ascorbic acid was 830% of the RDA for men and 570% of the RDA for women. The median supplemental intake for vitamin E was over 18 times the RDA; men and women taking supplements of all types had a median intake nearly 3 times the RDA.

The vitamin-supplement usage rate among the elderly has been reported as being between 35 and 75% of respondents (Dibble et al. 1967; LeBovit 1965; Steinkamp et al. 1965; Davidson et al. 1962; McGandy et al. 1966; Read and Graney 1982).

A 1982 Gallup poll study of vitamin-supplement use revealed that 37% of U.S. adults are users. Women are more likely to take vitamins than men. Usage increases with higher education and income. People living in the West use more vitamins. Lifestyle factors also indicate vitamin usage: nonsmokers, dieters, and those who exercise regularly are also more likely to be vitamin users. More than half of all U.S. vitamin users are either under 18 or over 59 years (MRCA 1981).

In the Nationwide Food Consumption Survey (NFCS) (1977–78) conducted by the U.S. Department of Agriculture, 34.7% of the 36,000 participants, six months of age or older, took supplements and 26.4% took them regularly. The highest regular use was found in infants less than one year of age.

The National Health and Nutrition Examination Surveys (NHANES I and NHANES II), carried out from 1971 to 1974 and from 1976 to 1980, respectively, asked approximately 20,000 participants about their use of vitamin and mineral supplements. In the NHANES I survey, 34% of the sample population reported use, compared with 37% in NHANES II. The highest use was in children from 1 to 5 years. Usage was also high among white females living above poverty income levels; in this group usage increased between the surveys. In the NFCS and the NHANES surveys, the most common type of supplement usage reported was of multiple-vitamin products with or without added minerals.

In the 1980 Food and Drug Administration (FDA) telephone survey, it was found that 39.9% of the sample took supplements regularly. Although the median level of supplement intake was approximately 200% of the RDA, 4.3% took vitamin A at levels of 25,000 IU per day, a potentially toxic dose. Heavy users of vitamin supplements received much of their information from health magazines and books. The most commonly consumed supplement was vitamin C (McDonald 1986).

REASONS FOR TAKING VITAMINS AND OTHER FOOD SUPPLEMENTS

Reasons cited for taking vitamins include "to stay healthy," "feeling run down," "not enough time for proper meals," and "illness" (Gallup Organization 1982). The 1978 A.C. Nielson survey showed that 51% took vitamins while on weight-reduction diets. Stanton (1983) provided evidence that people take vitamin supplements as a form of nutritional insurance. Specific vitamins are taken for their presumed health benefits. For example, vitamin C in amounts of 1 gram or more per day has been claimed to prevent and cure the common cold and other infections and to aid one's ability to withstand the effects of stress. Large doses of niacin have been advocated for use in the elderly to improve the circulation to the brain, preventing or lessening memory loss. Megadoses of thiamin are believed to relieve psychiatric disorders and prevent confusional states associated with senility (U.S. Dept. Agric. 1977–78).

It is fairly commonly held that elderly men and women take vitamins as well as other nutrient supplements because they accept media claims. Nonetheless, in the 1969 study of U.S. health practices and opinions carried out by the National Technical Information Administration (Rudman and Williams 1983), it was found that elderly vitamin users cited physicians and pharmacists as their main source of information on appropriate use of these supplements.

In 1976, Chaiton et al. also reported that vitamins are commonly prescribed under medical supervision. It is possible that the physician's assent to the patient's wishes is sometimes interpreted as both approval and recommendation.

Certain groups of physicians as well as osteopaths and chiropractors are adherents to "orthomolecular" practice. They believe that megadoses of specific vitamins will be taken up by the tissues of the body and converted to their active forms at a saturation level, where they will optimize metabolic processes (Hodges 1982).

Recent claims that nutrient supplements can prevent or cure cancer have also influenced vitamin-taking practices. An epidemiological association between low plasma vitamin A levels and increased risk of lung cancer was accompanied by scientific hypotheses based on epidemiological and animal studies indicating that vitamin A and beta-carotene may each play a role in cancer prevention. The popular press was not slow to overinterpret the facts as well as the theoretical proposals. A poster in a New York City pharmacy carried the message that you should take beta-carotene "to reverse the effects of smoking." Such misinformation may lead to inappropriate vitamin use.

Food-supplement usage includes not only vitamins but also minerals and trace elements. Mineral mixtures formulated and taken to supply physiological needs are increasingly popular. Other supplements include yeast, kelp, lecithin, garlic, bone meal, dolomite, fructose, proteins, amino acids, and formula foods.

PRESCRIPTION OF VITAMINS AND OTHER FOOD SUPPLEMENTS IN HOSPITALS, CLINICS, LONG-TERM CARE FACILITIES, AND MEDICAL OFFICES

Physicians may prescribe vitamins in order to supply the micronutrient needs of patients who are malnourished, who are on a marginal diet, or who are unable to take or retain normal food. Megadoses of B vitamins are correctly prescribed and administered orally or by parenteral routes to patients who have vitamin-dependent diseases including hereditary errors of metabolism as homocystinuria (Mudd and Levy 1983). Wernicke's encephalopathy, a thiamin-dependent syndrome in alcoholics during prolonged drinking sprees when they are not eating, is treated with B vitamin therapy (Wallace et al. 1978).

Therapeutic uses of micronutrients include niacin treatment of familial hypercholesterolemia, where the niacin is used for its hypolipidemic properties. Daily doses of niacin for these patients is on the order of 1 g/day.

Catheterized elderly as well as young paraplegic patients can be given vitamin C together with methenamine salts to acidify the urine and prevent urinary infections (Musher and Griffith 1974). However, it has been shown that the use of ascorbic acid to prevent infections in catheterized patients is without significant value (Nakarto et al. 1979).

Anorectic, debilitated, and depressed patients are sometimes given injections of vitamin B_{12} to promote appetite. Herbert (1977) has noted, however, that all objective studies have indicated that vitamin B_{12} has no appetite-stimulating properties and no effect on neurological disorders other than subacute combined degeneration occurring in pernicious anemia or other forms of vitamin B_{12} deficiency.

B vitamins may be prescribed for pregnant women who have had a previous child with a neural tube defect such as spina bifida. The justification is that in British epidemiological studies, associations have been observed between low B vitamin status of pregnant women and infants born with these malformations (Smithells et al. 1980).

Folic acid may be given in pharmacological doses to elderly institutionalized patients with dementia because of reports by several clinical investigators that patients with organic brain syndromes are likely to have inadequate folate status (Jensen and Olesen 1969; Kareks and Perry 1970). Sneath et al. (1973) were of the opinion that folate depletion in patients with dementias was related to an inadequate intake of dietary folate.

Folate supplementation has not been shown to cause significant clinical improvement in patients with organic brain syndromes. However, when Goodwin and Goodwin (1984) evaluated the association between nutritional status and cognitive functioning in 260 noninstitutionalized men and women over 60, individuals with low biochemical indices of vitamin C, vitamin B_{12}, folic acid, and riboflavin scored worse on the Halstead-Reitan categories test. This is a nonverbal, automated test of abstract thinking and problem-solving ability that is accepted as a sensitive indicator of minimal change in mental capability. The authors suggest that a prospective study should be carried out to examine the effects of vitamin supplementation on the cognitive status of elderly people.

PHARMACOLOGICAL USES OF VITAMINS AND VITAMIN ANALOGUES

Pharmacological uses of vitamins and vitamin analogues that are presently considered to be rational and appropriate are as follows:

1   The retinoid isotretinoin in the treatment of cystic acne (Peck et al. 1979).
2   The retinoid etretinate in the treatment of psoriasis (Wolska et al. 1983).
3   Niacin in the treatment of familial hyperlipoproteinemias (Havel and Kane 1982).
4   Thiamin in Wernicke's encephalopathy (Roe 1979).
5   Pyridoxine in the treatment of convulsions due to isoniazid (Brown 1972).

In addition, vitamins are used in high dosages in the treatment of congenital errors of metabolism:

1   Thiamin (100 mg/day) in the treatment of thiamin-responsive anemia and thiamin-dependent beriberi (Mandel et al. 1984).

TABLE 7.1 CLINICAL FEATURES OF HYPERVITAMINOSES: ACUTE AND CHRONIC EFFECTS, RISK FACTORS, AND TOXIC DOSE

Vitamin	Acute Effect of Pharmacological/ Toxic Dose	Symptoms & Signs of Chronic Vitamin Overload	Risk Factors	Toxic Dose (Chronic Day Dose)
Vitamin A	Nausea and vomiting Exfoliation	Dry skin, partial alopecia, loss of eyebrows, anorexia, bone pain, jaundice, hepatomegaly, ascites	Rx for cutaneous disease Viral or alcoholic liver disease	Infants ≥18,000 IU Normal Adults ≥50,000 IU Adults with liver disease ≥25,000 IU
Beta-Carotene	—	Carotenemia	Anorexia nervosa, erythropoietic protoporphyria	>30 mg
Vitamin D	Anorexia	Headache, confusion, polyuria, hypercalcemic anemia	Sarcoidosis	Infants >5,000 IU Adults ≥100,000 IU

110

Vitamin E	—	Intake of coumarin anticoagulants	Purpura	≥800 IU
Vitamin K	—	Prematurity	Hemolytic anemia, hyperbilirubinemia, kernicterus	>5 mg (water sol.)
Niacin	Flushing		Hepatic dysfunction, hyperglycemia, peptic ulcer, acanthosis nigricans	>1 g
Pyridoxine	—	Rx for carpal tunnel syndrome or premenstrual tension	Sensory neuropathy	>2 g
Vitamin C	—	Oxalate urolithiasis	Diarrhea, hyperoxaluria, renal calculi	>4 g
Folic Acid	—	Seizures on phenytoin	—	>5 mg

2 Pyridoxine (750 mg to 1.2 g/day) in homocystinuria and primary
   oxalosis (Barber and Spaeth 1967).
3 Folic acid (40 mg/day) in congenital disorders of folate absorption
   (Rowe 1966; Driskell 1984).

Epidemiological studies are in progress to evaluate the effectiveness
of high doses of beta-carotene as a cancer-protective agent (Wolf 1982).
The effectiveness of pharmacological doses of B vitamins given to pregnant women for the prevention of neural tube defects in their offspring
is being examined in intervention studies (Knox et al. 1983). Claims
that megavitamin therapy can benefit the hyperactive child have not
been substantiated, nor that vitamin C in megadoses is useful in the
treatment of cancer (Nutr. Rev. 1985; Pauling and Moertel 1986).

VITAMIN EXCESS

The toxic effects of vitamin excess are well recognized. According to
Rudman and Williams (1983), toxicity of the fat-soluble vitamins A
and D may be manifested at ten times the RDA, while megadoses of
water-soluble vitamins can be considered relatively harmless. Recent
information, however, indicates that high doses of some water-soluble
vitamins may be harmful. The clinical features of hypervitaminosis
are listed in Table 7.1; also listed are the risk factors and toxic dose
in previously normal people and in those who are especially vulnerable.

High intakes of carotene-rich foods or of pharmaceutical beta-carotene (Solatene, Roche) cause yellowing of the skin but not of the sclera.
This yellowing of the skin is usually designated carotenodermia or
aurantiasis. Carotenodermia is typically seen in diabetics on high
carotenoid intakes and patients with anorexia nervosa who may devour
raw carrots while excluding foods of higher caloric value. Daily doses
of 30–300 mg of beta-carotene are used in the photoprotection of patients with erythropoietic protoporphyria. The only consistent and significant side effect of this daily prophylactic treatment is the change
in skin color. Once carotenodermia has been established, it lasts from
2 to 6 weeks after cessation of treatment (Moore 1967).

Canthaxanthine (4,4-dioxy beta-carotene), a naturally occurring
carotenoid pigment, has also been produced synthetically. In the United
States it is used as a food coloring, while in Europe it serves in the
treatment of photosensitivity and as a tanning agent. Both canthaxanthine and beta-carotene in high doses can produce transient gastrointestinal symptoms, such as diarrhea (Matthews-Roth 1983).

HYPERVITAMINOSIS A

Hypervitaminosis A may be acute or chronic. Acute vitamin A overload may occur in young children who consume vitamin A liquid concentrates or capsules accidentally. It may also occur following consumption of polar bear liver. We assume the latter occurrence would be limited to Arctic explorers (Hayes and Hegsted 1973).

Clinical effects of acute retinoid toxicity from retinol, retinyl esters, or synthetic retinoids include headache, elevation of intracranial pressure, anorexia and nausea, cheilitis, dry mouth, fissures at the angles of the mouth, desquamation, alopecia, fatigue, vertigo, hepatospleno-megaly, and jaundice (Dicken 1984).

The synthetic retinoids are teratogenic. While it has long been recognized that teratogenic effects of pharmacological doses of retinol and retinoic acid can be produced in laboratory animals, a retinoic acid embryopathy, due to the synthetic retinoid isotretinoin, has been described in the infants of women who became pregnant while taking this drug orally. Major malformations seen in these infants include cleft palate, cardiac defects, thymic defects, optic nerve and retinal abnormalities, and malformations of the central nervous system (Lammer et al. 1985).

In the infant, signs of chronic hypervitaminosis include severe irritability, papilledema, and neurological features indicative of brain tumor (pseudotumor cerebri). Also, skeletal defects develop with irregular closure of the epiphyses of the long bones and resultant asymmetrical bone length (Ruby and Mital 1974). Enlargement of the liver and spleen as well as hypoplastic anemia, leukopenia, precocious skeletal development, and sparse, coarse scalp hair have been reported in a child who received approximately 240,000 IU of vitamin A per day from the age of three months to three years. These symptoms virtually cleared after the vitamin A "supplement" was discontinued (Moore 1967; Day 1978).

Chronic vitamin A toxicity in the adult is characterized by dryness of the skin, fissures of the corners of the mouth and nostrils, patchy alopecia with loss of eyebrows, liver dysfunction, bone pain, and increasing skeletal deformity (Muenter et al. 1971).

It has been stated that chronic symptoms of hypervitaminosis A occur with intakes of 25,000 to 50,000 IU/kg body weight (Bauernfeind 1980). However, toxicity varies with age, duration of intake, concurrent intake of natural and synthetic retinoids, nutritional status, as well as hepatic and possibly renal function. Vitamin A intoxication has occurred in infants receiving lower levels of vitamin A, and toxic intakes have been reported from 18,000 to 60,000 IU/day for infants less than one year old (Hayes and Hegsted 1973).

Hypervitaminosis A following prolonged low-level intake by an adolescent was reported by Farris and Erdman (1982). A 16-year-old boy had taken 50,000 IU of vitamin A for two and one-half years for acne. He had also taken other vitamins including vitamin E. Vitamin E may increase absorption of orally administered vitamin A (Arnrich 1978).

Hepatotoxicity developed in a 62-year-old man with protein-energy malnutrition, who ingested 40,000–50,000 IU of vitamin A per day for seven years (Weber et al. 1982). He had massive deposition of vitamin A in his liver (19,000 IU/g). Since the dosage was less than that commonly associated with vitamin A hepatotoxicity, and because there were no extrahepatic signs of vitamin A intoxication, it was postulated by the authors that the patient was unable to mobilize his hepatic stores of vitamin A because of protein deficiency. Laboratory data supported this hypothesis, in that the plasma level of retinol-binding protein doubled to a maximal value of 4.95 µg/dl after administration of a protein-adequate diet for 6 weeks.

A 42-year-old vegetarian male developed signs of hepatic encephalopathy (liver failure) secondary to viral hepatitis. He had been a moderate beer drinker. At the time of his hospital admission for management of the encephalopathy, he was observed to have generalized scaling and peeling of the skin. In view of the cutaneous and liver changes, and because he had already admitted taking vitamin C supplements, he was questioned about his intake of vitamin A. He stated that for ten years he had taken one 25,000 IU vitamin A capsule daily, and an additional 25,000 IU capsule when he was "under stress" or not feeling well. He ate liberal amounts of beta-carotene-containing foods, including carrot salad and leafy green vegetables. His estimated daily vitamin A intake was between 50,000 and 75,000 IU. Liver biopsy showed histological signs compatible with acute viral hepatitis. Lipid infiltration and periportal as well as perisinusoidal fibrosis were present. After recovery from hepatic encephalopathy, the patient developed universal alopecia which could have resulted from hypervitaminosis A (Hatoff et al. 1982).

Severe hepatotoxicity was reported in a 14-year-old girl following ingestion of vitamin A at levels of 100,000–200,000 IU/day for 15 months as treatment for eczema (Noseda et al. 1985). Her symptoms, which were of one month's duration, included shortness of breath and a complaint of abdominal swelling. She presented with a massive ascites, right-sided pleural effusion, and elevated serum alkaline phosphatase and gamma glutamyl transpeptidase levels. She was not icteric. At laparoscopy she was shown to have a large, pale liver with yellowish plaques. Liver biopsy showed fat accumulation in the perisinusoidal areas. Esophageal varices were demonstrated by endoscopy. No definite diagnosis was made at the time she first came under investigation, but one month later when the girl's mother reported her daughter's previous high intake of vitamin A, the serum value was found to be 150 μg/dl, four months after the young lady had discontinued intake of vitamin A supplements. At this time she admitted to hair loss previous to the initial hospital admission.

Hypervitaminosis A was diagnosed on the basis of electron microscopy study of liver biopsies that showed hypertrophied fat-storing cells in the spaces of Disse, lipofucsin granules, giant mitochondria with paracrystalline inclusions, and hypertrophy of the smooth endoplasmic reticulum. These changes were present at the time of her first hospital admission and also a month later. Though accumulation of an abnormal lipid material was observed in the liver tissue, liver vitamin A determinations were not made. Of particular concern was the fact that when the second liver biopsy was performed (one month after the first liver biopsy), there was an increase in the perisinusoidal collagen indicative of early liver fibrosis. The patient's pleural effusion and ascites resolved within six months after she was placed on a low vitamin A diet.

The intense pruritus (itching), dryness, and partial alopecia associated with end-stage renal disease may be related to high skin levels of vitamin A. In chronic renal failure, concentrations of vitamin A in the skin have been found similar to those of healthy controls (Vahlquist et al. 1982). Whether the cutaneous symptoms of patients with end-stage renal disease are exacerbated by intake of high doses of vitamin A is presently unknown.

SYNTHETIC RETINOIDS RELATED TO VITAMIN A

The synthetic retinoid, isotretinoin, the 13-cis isomer of retinoic acid, is in therapeutic use for the control of acne vulgaris and other acneiform conditions such as acne rosacea (Cunningham and Ehmann 1983). It is also used in the management of disorders of keratinization (Randle et al. 1980; Baden et al. 1982).

The recommended oral dose of isotretinoin for the treatment of acne is 1–2 mg/kg for a period of 15 to 20 weeks. The usual dose for the treatment of ichthyosiform dermatosis and other errors of keratinization is 2–3 mg/kg for periods of a year or more.

Side effects of this and other synthetic retinoids bear many similarities to those of hypervitaminosis A. A comparison of the toxic effects

TABLE 7.2 COMPARISON OF MAJOR TOXIC EFFECTS OF VITAMIN A
AND THE SYNTHETIC RETINOID ISOTRETINOIN

	Vitamin A	Isotretinoin
Skin	Dryness	Xerosis (dry skin)
	Nasal fissures	Desquamation
	Pruritus	Nasal fissures
	Jaundice	Pruritus
Hair	Thinning	Thinning (rare)
	Alopecia	
Mucosae		Conjunctivitis
		Cheilitis (dry lips)
		Xerostomia (dry mouth)
		Epistaxis
Neurological	Headache	Headache
	Pseudotumor cerebri	
	(infants)	
Gastrointestinal	Hepatomegaly	
Musculoskeletal	Bone/joint pain	Bone/joint pain
	Hyperostoses	Hyperostoses
	Premature	Premature epiphyseal closure
	epiphyseal closure	
Pregnancy	Birth defects	Birth defects
	Abnormal liver	Abnormal liver function tests
		(rare)
Lipid profile		Elevated triglycerides

of vitamin A and this retinoid is shown in Table 7.2. At the therapeutic drug level used in the treatment of acne, the main side effects are dryness of skin with reduced secretion of sebum, desquamation, cheilitis, and conjunctivitis. Elevated serum triglyceride levels with synthetic retinoids are more common in patients who ingest alcohol regularly, in obese individuals, and in patients with diabetes mellitus. Hypertriglyceridemia is more common with higher doses of these retinoids for the treatment of ichthyosiform dermatosis, where 25% of patients may be affected (Windhorst and Nigra 1982).

Although there is little hepatic storage of isotretinoin, abnormal liver function tests occur in about 10% of patients receiving ≥ 2 mg/kg of the drug (Cunningham and Ehmann 1983; Brazzell and Colburn 1982). The drug is 99% bound to plasma albumin; it is probable that blood levels are influenced by plasma albumin levels. It is strongly recommended that *all* patients receiving synthetic retinoids abstain from taking vitamin A supplements and that patients who are taking vitamin A supplements in doses greater than 5,000 IU/day not receive isotretinoin until at least three months have elapsed. Certain effects of hypervitaminosis A and the synthetic retinoids may be additive.

Hypercalcemia has been described with megadose intake of vitamin A, and also as a side effect of isotretinoin. This complication of retinoid therapy developed in a 19-year-old male with cystic acne who received the drug at a daily dose of 1.6 mg/kg. The hypercalcemia was associated with bone pain, nausea, lethargy, and anemia (Valentie et al. 1983). It is believed that retinoids, including vitamin A, cause accelerated mineral resorption both in trabecular and in cortical bone as a result of increased osteoclastic activity, and that retinoids potentiate release of lysosomal enzymes involved in bone resorption (Frame et al. 1974).

HYPERVITAMINOSIS D

Prolonged excessive intake of vitamin D can result in hypercalcemia and metastatic, extraosseous calcification. The calcium deposits occur particularly in arterial walls and in the kidney. Tuberous periarticular deposits of calcium may be seen. There is progressive impairment of renal function leading to renal failure. The signs and symptoms of hypervitaminosis D in the child or adult include anorexia, nausea, vomiting, thirst, constipation, sometimes fever, abdominal pain, pallor, fatigue, diarrhea, and polyuria (Taussig 1966). In the elderly, hypervitaminosis D can produce an organic brain syndrome (Verner et al. 1958).

An "epidemic" of infantile hypercalcemia occurred in the United Kingdom from 1952 to 1954. The infants, usually from three to nine months, presented with failure to thrive, irritability, vomiting, and in some cases, evidence of renal damage. The problem was traced to concurrent intake of cod liver oil and vitamin D–fortified reconstituted dried milk (Wharton and Darke 1982; Lightwood 1952).

Evidence shows that in vitamin D intoxication, the toxic form of the vitamin is 25-hydroxycholecalciferol (Haussler and McCain 1977). It has been reported that hypercalcemia can occur following the administration of alpha-hydroxycholecalciferol (Lund et al. 1975). The toxic dose of vitamin D shows marked interindividual variability (Parfitt et al. 1982).

VITAMIN E

In the adult, megadoses of vitamin E have been shown to potentiate the effects of coumarin anticoagulants and therefore to increase the risk of hemorrhage (Corrigan and Marcus 1974). The suggestion has been made by Arnrich (1978) that metabolites of alpha-tocopherol, such as alpha-tocopheryl quinone or hydroquinone, may act as antimetabolites to vitamin K.

A group of Swedish investigators have reported a prolonged plasma clotting time in human subjects after long-term treatment with alpha-tocopherol (Korsan-Bengsten et al. 1974). Corrigan and Ulfers (1981) studied the role of vitamin E in blood coagulation. They found that activity of the precursor of coagulation factor II (prothrombin) is reduced in warfarin-treated animals and humans given vitamin E.

In neonatal infants given vitamin E as an intramuscular injection, extensive ossification has been reported at the site of injection in the thigh. The vitamin E was administered to these infants for the prevention of retrolental fibroplasia at a daily dose of 25 mg/kg (Smith et al. 1983). In one clinical trial of vitamin E in premature low-birth-weight infants, the incidence of necrotizing enterocolitis and sepsis was increased (Hiltner et al. 1983).

VITAMIN K

Water-soluble preparations of vitamin K_3 (menadione) are toxic to premature infants during the neonatal period when administered by injections. Hemolytic anemia results from oxidative damage to red cells, and consequently the immature liver floods with bilirubin derived from heme. Passage of bilirubin to the basal nuclei of the brain results in hyperbilirubinemia as well as kernicterus. Kernicterus causes brain damage, which can lead to cerebral palsy if the infant survives (Nutr. Rev. 1961).

VITAMIN C

Moderate hyperoxaluria has been induced in normal men by megadoses of vitamin C (≥ 4 g daily) given for seven days. In children and adults with oxalate urolithiasis (predisposition to form oxalate kidney stones), it has been shown that ascorbic acid ingestion at a dose of 2 g/day can further increase an already elevated oxalate excretion (Takenouchi et al. 1966; Briggs et al. 1973).

In patients with renal failure who are receiving hemodialysis, intravenous ascorbic acid causes increases in plasma oxalate levels and deposition of oxalate crystals in the kidneys, often leading to an adverse effect on prognosis (Balcke et al. 1984).

Vitamin C in large doses does diminish the anticoagulant effect of heparin and of coumarin anticoagulants (Rosenthal 1971; Owen et al. 1970). Megadoses of ascorbic acid can produce cramping, abdominal pain, and diarrhea, particularly when the vitamin is taken before meals (Korner and Weber 1972).

VITAMIN B_6

Recent claims for the therapeutic value of vitamin B_6 together with general acceptance of pyridoxine as a safe vitamin provide reasonable confidence in megadose usage. Tablets containing 50–500 mg of pyridoxine are generally available. It is claimed to be of benefit in the treatment of carpal tunnel syndrome in doses on the order of 300 mg/day. Schizophrenic adults and autistic children have been given doses from 600–3,000 mg/day. Doses up to 2 g/day have been given to hyperactive children (Ellis et al. 1981; Pauling et al. 1973; Rimland et al. 1978).

However, pyridoxine in very high doses has been shown to be toxic to animals and to man. The subacute toxicity of pyridoxine hydrochloride was studied in beagle dogs (Phillips et al. 1978). Clinical signs of neurotoxicity developed between 40 to 75 days after the onset of treatment in a group of dogs receiving 200 mg/kg/day. Signs included ataxia and muscle weakness. Histopathological examination showed bilateral loss of myelin and axons in the dorsal funiculi of the spinal cord and myelin loss of nerve fiber of the dorsal nerve roots.

A sensory neuropathy has also been described in five women and two men taking 2–6 g of pyridoxine/day for a period of from 2 to 40 months. The patients displayed a progressive sensory ataxia, impairment of position and vibration sense in the limbs, and diminished or absent tendon reflexes. Sural nerve biopsy revealed nerve fiber degeneration and nerve loss. When the vitamin was withdrawn, slow and partial recovery resulted (Schaumburg et al. 1983).

An editorial on this report of megavitamin abuse pointed out that pyridoxine is a substituted pyridine and that pyridines are neurotoxic. However, there was some question whether the actual neurotoxin was pyridoxine or an impurity in the formulated product due to degradation (U.S. Dept. Agric. 1977–78). In view of the findings in the studies of the beagles, it would seem that the pyridoxine was the toxic substance.

Skin changes, including blisters, have been reported in an alcoholic woman who took 4 g of pyridoxine daily for four years. She also presented with neuropathy (Baer and Stillman 1984).

Lower levels of pyridoxine intake by individuals whose pyridoxine dosage varied from 200 mg to 5 g/day was associated with symmetric, distal sensory loss. In four of the patients, vibration and position sense were disproportionately reduced. Sensory ataxia was present in seven patients. Sensory nerve action potentials were absent or severely reduced in this patient group. Follow-up information on these individuals, 3 to 18 months after their pyridoxine intake was reduced, indicated symptomatic improvement but incomplete recovery (Parry and Bredesen 1985).

Other than in the management of genetically determined vitamin B_6 dependency syndromes, the only condition in which high doses of vitamin B_6 should be recommended is in the treatment of acute isoniazid intoxication. Isoniazid (INH) is a vitamin B_6 antagonist that in massive doses produces seizures. When pyridoxine was given with or without anticonvulsants as the antidotal drug to dogs that previously received convulsant doses of INH, neurotoxicity of the vitamin was not observed (Chin et al. 1981). In INH poisoning, it has been recommended that 1 g of pyridoxine be given for each gram of INH estimated to have been ingested by the patient (Brown 1972).

NIACIN

Because pharmacological doses of niacin produce flushing, this vitamin has been widely used as a vasodilator drug (Coffman 1979). Ruffin and Smith (1939) noted that in people receiving 250 mg niacin 4 times/day, additional symptoms included lassitude, depression, palpitations, cyanosis, substernal discomfort, headache, nausea, vomiting, and dyspnea. These effects have not been reproduced by subsequent investigators. However, it is possible that their observations were caused by histamine release from niacin and/or from a contaminant. Prostaglandin releasers, including aspirin and indomethacin, block the niacin flush reaction (Wilkin et al. 1981).

Because niacin liberates histamine and thus triggers the release of hydrochloric acid in the stomach, it has long been surmised that large doses of the vitamin would aggravate peptic ulcer disease (PUD). PUD has been reported in 9 out of 68 patients taking 3–7.5 g/day of niacin for 96 to 130 weeks (Mosher 1970).

Niacin has been used to treat hyperlipemias. The niacin may be administered as a single "drug" or in combination with another hypolipidemic agent, such as cholestyramine. In pharmacologic doses, niacin acts to inhibit the production of very low-density lipoproteins (VLDL) and the daughter low-density lipoproteins (LDL). It is also known to increase fecal cholesterol excretion and inhibit hepatic cholesterol biosynthesis. Another effect that may be beneficial in the treatment of patients with atherosclerosis is that niacin inhibits the catabolism of high-density lipoproteins (HDL) (Havel and Kane 1982).

As a hypolipidemic drug, the daily dose of niacin is on the order of

1–3 g per day. Patients receiving niacin for the treatment of hyperlip-idemia usually begin to experience flushing when they are taking about 200–300 mg/day (Kane et al. 1981). At higher doses other signs of toxicity are manifested, including hepatotoxicity associated with ele-vated serum transaminase levels. Glucose intolerance and hyperuri-cemia occur quite frequently. Rarer complications include cardiac ar-rhythmias, toxic amblyopia, and development of the pigmentary dermatosis acanthosis nigricans (Roe 1966).

The hepatotoxicity of niacin was first reported in dogs in 1938, one year after the discovery that nicotinic acid (niacin) is the pellagra-preventing vitamin. In that report, a dog weighing 13.8 kg developed a fatty liver after receiving a dose of 2 g of the vitamin (Chen et al. 1938). High levels of niacin have been associated with development of cholestatic jaundice (Einstein et al. 1975). In reporting on the recorded cases of hepatotoxicity due to niacin, Alhadeff et al. (1984) emphasized that the dose causing liver changes may be quite variable. For example, in one report 750 mg of the vitamin per day for three months induced cholestasis (Sugerman and Clark 1974). However, a single case of he-patic necrosis was documented in a woman who took 3 g of the vitamin per day for several years (Alhadeff et al. 1984).

In the last ten years, niacin has been used by orthomolecular psy-chiatrists at doses up to and exceeding 6 g/day in the treatment of schizophrenia.

RISKS OF MULTIVITAMIN PREPARATIONS

Multivitamin preparations may contain both water- and fat-soluble vitamins. It is not uncommon for individuals who are vitamin abusers to take multiple multivitamin preparations that then impose the risk of multiple hypervitaminoses (Alhadeff et al. 1984). See the adverse effects listed in Table 7.3.

MEGAVITAMIN INTAKE DECREASES THE EFFECTIVENESS OF PRESCRIPTION DRUGS

Although people who have seizure disorders and are taking the drug phenytoin (Dilantin) and/or phenobarbital regularly have an increased requirement for the B vitamin folic acid, seizure control may be lost with high intake of this vitamin. Intravenous folate can actually pre-cipitate fits (Reynolds 1973; Chanarin et al. 1960; Chien et al. 1975).

Supplementary folate in low doses (0.1–1 mg daily) throughout preg-nancy is sufficient to prevent folate deficiency even in women with seizure disorders who are receiving anticonvulsant drugs. This dose does not impair seizure control as can higher doses of folic acid (5 mg

TABLE 7.3 ADVERSE EFFECTS OF MEGADOSE VITAMIN ADMINISTRATION

(Fat-Soluble Vitamins) Vitamin Analog	Adverse Effect	Daily Dose	Duration of Usage	Predisposing Factors
Vitamin D	*Infants* Hypercalcemia Irritability	≥4000 IU	>3 months	
	Adults Headache Hypercalcemia Soft tissue calcification	≥25,000 IU	>3 months	Sarcoidosis
	Renal impairment Uremia			Renal disease
Vitamin E	Coagulopathy	≥800 IU	>4 weeks	Intake of Warfarin anticoagulant
Vitamin K (Vitamin K₃)	*Infants* Hemolytic anemia	>5 mg to mother >1 mg to infant	1 day	Prematurity
	Hyperbilirubinemia			
	Kernicterus			
Beta-carotene	Aurantiasis (carotenemia)	≥60 mg	2 weeks	Anorexia nervosa EPP*
Vitamin A	*Prenatal* Birth defects	≥50,000 IU (to mother)	>1 week	Pregnancy with prior intake oral contraceptives
	Infants Pseudotumor cerebri Premature epiphyseal closure	≥25,000	≥3 months	

	Adverse Effect	Daily Dose	Duration	Predisposing Factors
Adults Alopecia Desquamation Hyperostoses Hypercalcemia Hepatic injury		>50,000 IU >90,000 IU	≥1 year ≥1 year	Renal disease Alcoholic/viral hepatitis Protein deficiency
13-CIS retinoic acid	*Prenatal* Birth defects	≥1 mg/Kg (to mother)	unknown	Pregnancy with intake of vitamin A
	Adults Desquamation Hyperostoses Hypercalcemia Hyperlipemia	≥1 mg/Kg ≥3–4 mg/Kg	2–6 years	Intake of vitamin A

(Water-Soluble Vitamins) Vitamin	Adverse Effect	Daily Dose	Duration	Predisposing Factors
Pyridoxine (Vitamin B₆)	Sensory neuropathy	2–6 g	2–40 months	Premenstrual tension Carpal tunnel syndrome
Niacin	Flushing	≥200 mg	15–30 min	Raynaud's disease Hyperlipemic state Schizophrenia
	Hepatic dysfunction Glucose intolerance	>3 g	≥1 year	
Folic acid	Loss of seizure control	>5 mg	1 day	Epilepsy—on phenytoin
Ascorbic acid	Diarrhea Renal calculi	>500n mg	1 week ≥1 year	Primary oxalosis

*EPP = erythropoietic protoporphyria

of folic acid or more) (Hilesmaa et al. 1983; Reynolds et al. 1972). Vitamin B_6 can also interfere with seizure control.

Vitamin K supplementation from liquid formula products has been shown to interfere with the efficacy of coumarin anticoagulants. Resistance to the anticoagulant drug warfarin from this cause is now less of a risk than formerly because manufacturers have decreased the vitamin K content of formula diets and supplements (O'Reilly and Rytand 1981; Lee et al. 1981).

Riboflavin has been shown to inhibit the uptake of the cancer chemotherapy drug, methotrexate, into malignant cells (DiPalma and Ritchie 1977).

MINERAL AND TRACE ELEMENT TOXICITY FROM FOOD SUPPLEMENTS, FOOD ADDITIVES, AND FOOD CONTAMINANTS

Toxic hazards from mineral and trace element supplements include acute poisoning, chronic toxic effects from overload, and heavy metal poisoning due to product contamination.

The acute and chronic toxicity of several minerals in supplements is well described. Readers are advised to review the information given in the 1978 Report of the Ad Hoc Committee, appointed by the National Nutrition Consortium to examine potential hazards of excess consumption (Report of Ad Hoc Committee 1978).

Acute iron toxicity is still a serious hazard in young children who may ingest ferrous sulfate or other iron pills intended for a parent or older sibling (Robinson 1975).

In normal people, iodine overload is not caused by consuming an ordinary mixed diet, but may occur when food or beverages contain excessive amounts of iodine from contamination. Milk can be contaminated by iodophors used to clean dairy machinery and from use of iodine-containing feed additives. Although iodine consumption from breakfast cereals containing tetraiodofluorescein as a red food dye has been documented, the health risks are not well defined. Iodide goiter solely from ingestion of high-iodine milk and cereals has not been reported (Ferri 1985).

Iodine excess leading to the development of iodide goiter and iodide-induced thyrotoxicosis may result from the ingestion of kelp (Fradkin and Wolff 1983). We have seen a young woman who had a viral thyroiditis and then developed thyrotoxicosis after taking kelp daily for 18 months.

In individuals with impaired renal function, magnesium-containing mineral supplements may cause a magnesium overload (Randall et al. 1964). Calcium supplements may cause a decline in renal function in individuals who already have renal disease and increased blood levels of phosphate (Walser 1980). Calcium carbonate supplements taken at bedtime can reduce dietary iron absorption. High doses of calcium carbonate (>1 g/day), presently being recommended for the prevention of osteoporosis and hypertension, can cause the milk alkali syndrome and hypercalcemia, particularly in individuals with reduced renal function (Orwoll 1982; Ann. Intern. Med. 1985).

OVERUSE OF VITAMINS AND MINERALS

The prevalence of nutrient supplement use in the United States illustrates the public's belief that vitamins and minerals can prevent disease, provide nutritional insurance, and change destiny. Physicians prescribe vitamins or minerals for patients at nutritional risk and for those whose diets are adequate. Megadoses of vitamins have been claimed not only to prevent the common cold but also to control schizophrenia. Vitamin–mineral supplements are justified in the prevention of deficiencies in those eating marginal diets, and in the treatment of specific nutritional deficiencies. Vitamins (e.g., beta-carotene) are proposed for cancer prevention.

Since vitamin and mineral supplement usage is presently in excess of demonstrated need, and since megadoses of these nutrients may be toxic as well as cause interference with therapeutic drugs, it is recommended that the lay public be reminded of the potential dangers of self-medication with high doses of vitamins and minerals. The information provided to the public should stress that there are few indications for megavitamin dosage and that these indications should be identified by a physician. Further, the public needs to be protected by limiting of the unit amounts of any vitamin or mineral that can be purchased, and by patient package inserts that clearly define the dangers of overdosage.

Various attempts have been made to explain the extent of food supplement misuse and abuse. Blame has been laid on promotion by the health food industry, by pharmaceutical companies, and by drugstores (Jarvis 1983). Megadose vitamin therapy prescribed by physicians has been criticized because clinical testimonials advanced to support this practice are unacceptable today in an era of extensive data collection and clinical trials (DiPalma and McMichael 1982).

No matter how much we lay the blame on industry or on health professionals for encouraging excessive use of food supplements, nutrition professionals need to take more responsibility for guiding the lay public in appropriate use of vitamins and minerals. It is important that individuals who are particularly at risk for vitamin or mineral overload be warned against taking supplements that, for them, can lead to nutrient excess.

REFERENCES

ALHADEFF, L., C.T. GUALTIERI, and M. LIPTON. 1984. Toxic effects of water soluble vitamins. *Nutr. Rev.* 42: 33–39.

ANNALS OF INTERNAL MEDICINE. 1985. Editorial. Calcium tablets for hypertension? *Ann. Intern. Med.* 103: 946–47.

ARNRICH, L. 1978. Toxic effects of megadoses of fat-soluble vitamins. In *Nutrition and Drug Interrelations*. ed. J.N. Hathcock and J. Coon. 751–71. New York: Academic Press.

BADEN, H.P., M.M. BUXMAN, G.D. WEINSTEIN, and E.W. YODER. 1982. Treatment of ichthyosis with isotretinoin. *J. Am. Acad. Dermatol.* 6: 716–20.

BAER, R.L. and M.A. STILLMAN. 1984. Cutaneous skin changes probably due to pyridoxine abuse. *J. Am. Acad. Dermatol.* 10:527–28.

BALCKE, P., P. SCHMIDT, J. ZAZGORNIK, H. KOPSA, and A. HAUENSTOCK. 1984. Ascorbic acid aggravates secondary hyperoxalemia in patients on chronic hemodialysis. *Ann. Intern. Med.* 101: 344–45.

BARBER, G.W. and F.L. SPAETH. 1967. Pyridoxine therapy in homocystinuria. *Lancet* 1: 337.

BAUERNFEIND, J.C. 1980. *The safe use of vitamin A.* A report of the international vitamin A consultative group. The Nutrition Foundation.

BRAZZELL, R.K. and W.A. COLBURN. 1982. Pharmacokinetics of the retinoids isotretinoin and etretinate. *J. Am. Acad. Dermatol.* 6: 643–51.

BRIGGS, M.H., P. GARCIA WEBB, and P. DAVIES. 1973. Urinary oxalate and vitamin C supplements. *Lancet* 2: 201.

BROWN, C.V. 1972. Acute isoniazid poisoning. *Am. Rev. Resp. Dis.* 105: 206–16.

CHAITON, A., W.O. SPITZER, R.S. ROBERTS, and T. DELMORE. 1976. Patterns of medical drug use—a community focus. *Can. Med. Assoc. J.* 14: 33–37.

CHANARIN, I., J. LAIDLAW, L.W. LOUGHBRIDGE, and D.L. MOLLIN. 1960. Megaloblastic anemia due to phenobarbitone: the convulsant action of therapeutic doses of folic acid. *Brit. Med. J.* 1: 1099.

CHEN, K.K., C.L. ROSE, and E. BROWN ROBBINS. 1938. Toxicity of nicotinic acid. *Proc. Soc. Exp. Biol. Med.* 38: 241–45.

CHIEN, L.T., C.L. KRUMDIECK, C.W. SCOTT, Jr., and C.E. BUTTERWORTH, Jr. 1975. Harmful effect of megadoses of vitamins: electroencephalogram abnormalities and seizures induced by intravenous folate in drug-treated epileptics. *A.J.C.N.* 28: 51.

127

CHIN, L., M.L. SIEVERS, R.N. HERNER, and A.L. PICCHIONE. 1981. Potential of pyridoxine by depressants and anticonvulsants in the treatment of acute isoniazid intoxication in dogs. *Tox. Appl. Pharmacol.* 58: 504–9.

COFFMAN, J.D. 1979. Vasodilator drugs in peripheral vascular disease. *New Engl. J. Med.* 300: 713–17.

CORRIGAN, J.J. Jr. and F.I. MARCUS. 1974. Coagulopathy associated with vitamin E ingestion. *J.A.M.A.* 230: 1300–1301.

CORRIGAN, J.J. Jr. and L.L. ULFERS. 1981. Effect of vitamin E on prothrombin levels in warfarin-induced vitamin K deficiency. *A.J.C.N.* 34: 1701–4.

CUNNINGHAM, W.J. and C.W. EHMANN. 1983. Clinical aspects of the retinoids. *Seminars in Dermatology* 2: 145–60.

DAVIDSON, C.S., J. LIVERMORE, R. ANDERSON, and S. KAUFMAN. 1962. The nutrition of a group of apparently healthy aging persons. *A.J.C.N.* 10:181–99.

DAY, H.G. 1978. Vitamin A. In *Vitamin-Mineral Safety, Toxicity and Misuse.* 2–4. National Nutrition Consortium, Inc., Chicago: A.D.A.

DIBBLE, M.V., M. BRIN, V.F. THIELE, A. PEELE, N. CHEN, and E. McMULLEN. 1967. Evaluation of nutritional status of elderly subjects with a comparison between fall and spring. *J. Am. Geriat. Soc.* 15: 1031–61.

DICKEN, C.H. 1984. Retinoids: a review. *J. Am. Acad. Dermatol.* 11: 541–52.

DIPALMA, J.R. and R. McMICHAEL. 1982. Assessing the value of meganutrients in disease. *Bull. N.Y. Acad. Med.* 58: 254–62.

DIPALMA, J.R. and D.M. RITCHIE. 1977. Vitamin toxicity. *Ann. Rev. Pharm. Tox.* 17: 133.

DRISKELL, J.A. 1984. Vitamin B_6. In *Handbook of Vitamins.* ed. L.J. Machlin. 379–401. New York: Marcel Dekker.

DUBICK, M.A. and R.B. RUCKER. 1983. Dietary supplements and health aids—a critical evaluation. Part I, Vitamins and Minerals. *J. Nutr. Educ.* 15: 47–53.

EINSTEIN, N., A. BAKER, J. GALPER, and H. WOLFE. 1975. Jaundice due to nicotinic acid therapy. *Am. J. Dig. Dis.* 20: 282–86.

ELLIS, J., K. FOLKERS, M. LEVY, K. TAKEMURA, S. SHIZUKUISHI, R. ULRICH, and P. HARRISON. 1981. Therapy with vitamin B_6 with and without surgery for treatment of patients having the idiopathic carpal tunnel syndrome. *Res. Commun. Chem. Pathol. Pharmacol.* 33: 331–44.

FARRIS, W.A. and J.W. ERDMAN. 1982. Protracted hypervitaminosis A following long-term low-level intake. *J.A.M.A.* 247: 1317–18.

FERRI, G. 1985. The effect of excess dietary iodine on thyroid function in New York State dairy farm families. M.S. Thesis, Cornell University, Ithaca, NY.

FRADKIN, J.E. and J. WOLFF. 1983. Iodide-induced thyrotoxicosis. *Medicine* 62: 1–20.

FRAME, B., C.E. JACKSON, W.A. REYNOLDS, and J.E. UMPHREY. 1974. Hypercalcemia and skeletal effects of chronic hypervitaminosis A. *Ann. Intern. Med.* 80: 44–48.

GAGNON, J.P., E.J. SALBER, and S.B. GREENE. 1978. Patterns of prescription and non-prescription drug use in a southern rural area. *Public Health Rep.* 93: 4433–54.

GALLUP ORGANIZATION. 1982. *The Gallup study of vitamin use in the United States.* Survey 6.6. Princeton, NJ: Gallup.

GARRY, P.J., J.S. GOODWIN, W.C. HUNT, E.M. HOOPER, and A.G. LEONARD. 1982. Nutritional status in a healthy elderly population: Dietary and supplemental intakes. *A.J.C.N.* 36: 319–31.

GOODWIN, J.S. and J.M. GOODWIN. 1984. Association between nutritional status and cognitive functioning in a healthy elderly population. *J.A.M.A.* 249: 2917–40.

HATOFF, D.E., S.L. GERTLER, K. MIYAI, B.A. PARKER, and J.B. WEISS. 1982. Hypervitaminosis A unmasked by acute viral hepatitis. *Gastroenterology* 82: 124–28.

HAUSSLER, M.R. and T.A. McCAIN. 1977. Basic and clinical concepts related to vitamin D metabolism and action, 2nd of 2 parts. *New Engl. J. Med.* 297:1041–50.

HAVEL, R.J. and J.P. KANE. 1982. Therapy in hyperlipidemic states. *Ann. Rev. Med.* 33: 417–33.

HAYES, K.C. and D.M. HEGSTED. 1973. Toxicity of the vitamins. In *Toxicants Occurring Naturally in Foods.* Committee on Food Protection. 2nd ed. Washington: National Academy of Sciences.

HERBERT, V. 1977. Megavitamin therapy. In *Contemporary Nutrition 2,* Minneapolis, MN: General Mills, Inc., Nutrition Department.

HILESMAA, V.K., K. TERAMO, M.L. GRANSTROM, and A.H. BARDY. 1983. Serum folate concentrations during pregnancy in women with epilepsy: relation to antiepileptic drug concentrations, number of seizures and fetal outcome. *Brit. Med. J.* 287: 577–79.

HILTNER, H.M., F.L. KRETZER, A.J. RUDOLPH, H.E.B. HOLBEIN, G. TROENDLE, and S. SOBEL. 1983. Vitamin E in retrolental fibroplasia. *New Engl. J. Med.* 309: 669–70.

HODGES, R.E. 1982. Megavitamin therapy. *Primary Care* 9: 605–19.

JARVIS, W.T. 1983. Food faddism, cultism and quackery. *Ann. Rev. Nutr.* 3: 35–52.

JENSEN, O.N. and O.V. OLESON. 1969. Folic acid concentrations in psychiatric patients. *Acta Psychol. Scand.* 45: 289–94.

KANE, J.P., M.J. MALLEY, P. TUN, N.R. PHILLIPS, D.D. FREEDMAN, M.L. WILLIAMS, J.S. ROWE, and R.J. HAVEL. 1981. Normalization of low-density lipoprotein levels in heterozygous familial hypercholesterolemia with a combined drug regimen. *New Engl. J. Med.* 304: 251–57.

KAREKS, J. and S.W. PERRY. 1970. Folic acid deficiency in psychiatric patients. *Med. J. Aust.* 1: 1192–95.

KNOX, E.G., I. EMANUEL, and R.W. SMITHELLS. 1983. Vitamin supplementation and neural tube defects. *Lancet* 2: 39.

KORNER, W.F. and F. WEBER. 1972. Tolerance of high ascorbic acid doses. *Internat. J. Vitamin Res.* 42: 528.

KORSAN-BENGSTEN, K., D. ELMFELDT, and T. HOLM. 1974. Prolonged plasma clotting time and decreased fibrinolysis after long term treatment with alpha-tocopherol. *Throm. Diath. Haemorrh.* 31: 505–12.

LAMMER, E.J., D.T. CHEN, R.M. HOAR, N.D. AGNISH, P.J. BENKE, J.T. BRAUN, C.J. CURRY, P.M. FERNHOFF, A.W. GRIX, I.T. LOTT, J.M. RICHARD, and S.C. SUN. 1985. Retinoic acid embryopathy. *New Engl. J. Med.* 313: 837–41.

LAMY, P.P. 1982. Effects of diet and nutrition on drug therapy. *J. Am. Geriat. Soc.* 30: S99–S111.

LEBOVIT, C. 1965. The food of older persons living at home. *J.A.D.A.* 46: 285–89.

LEE, M., R.N. SCHWARTZ, and R. SHARIFI. 1981. Warfarin resistance and vitamin K. *Ann. Intern. Med.* 94: 140–41.

LIGHTWOOD, R.C. 1952. Idiopathic hypercalcaemia in infants with failure to thrive. *Arch. Dis. Child.* 27: 302.

LUND, B., L. HJORTH, J. KJAER, N. RAMAN, T. FRIIS, R.B. ANDERSON, and D.H. SORENSEN. 1975. Treatment of osteoporosis of aging with 1-alpha-hydroxycholecalciferol. *Lancet* 2: 1168–71.

MANDEL, H., M. BERANT, A. HAZANI, and Y. NAVEH. 1984. Thiamine-dependent beriberi in the thiamine-responsive anemia syndrome. *New Engl. J. Med.* 311: 836–38.

MATTHEWS-ROTH, M.M. 1983. Anthraxanthen-produced skin tan: Carotenodermia by any other name. *J.A.M.A.* 250: 1100.

McDONALD, J.T. 1986. Vitamin and mineral supplement use in the United States. *Clin. Nutr.* 5: 27–33.

McGANDY, R.B., C.H. BARROWS, A. SPAMAS, A. MEREDITH, J.L. STRONG, and A. NORRIS. 1966. Nutrient intakes and energy expenditure in men of different ages. *J. Gerontol.* 21: 581–87.

MOORE, T. 1967. Pharmacology and toxicity of vitamin A. In *The Vitamins.* 2nd ed. ed. W.H. Sebrell and R.S. Harris. 280–94. New York: Academic Press.

MOSHER, L.R. 1970. Nicotinic acid side effects and toxicity; a review. *Am. J. Psych.* 126: 1290–96.

MRCA QUARTERLY REPORT. 1981. Market Research Corp. of America, Northbrook, IL.

MUDD, S.H. and H.L. LEVY. 1983. Disorders of trans-sulfuration. In *The Metabolic Basis of Inherited Disease.* 5th ed. ed. J.B. Stanbury, J.B. Wyngaarden, D.S. Fredrickson, J.L. Goldstein, and M.S. Brown. 522–59. New York: McGraw-Hill.

MUENTER, M.D., H.O. PERRY, and J. LUDWIG. 1971. Chronic vitamin A intoxication in adults. *Am. J. Med.* 30: 1291.

MUSHER, D.M. and D.P. GRIFFITH. 1974. Generation of formaldehyde from methenamine: effect of pH and concentration and antibacterial effect. *Antimicrob. Agents and Chemother.* 6: 708–11.

NAKARTO, D.V., C.J. BELL, and P.P. LAMY. 1979. Appraisal of ascorbic acid for acidifying the urine of methenamine-treated geriatric patients. *J. Am. Geriat. Soc.* 27: 34–37.

NATIONWIDE FOOD CONSUMPTION SURVEY. 1981. *Food and Nutrient Intakes of Individuals in One Day in the U.S., 1977–78.* U.S.D.A. Consumer Nutrition Center Preliminary Report no. 2. Washington, D.C.: U.S. Gov't Ptg. Office.

NATIONWIDE FOOD CONSUMPTION SURVEY. 1983. *Food Consumption in Households in the U.S.: Seasons and Years, 1977–78.* U.S.D.A. Human Nutrition Information Service no. 6. Washington, D.C.: U.S. Gov't Ptg. Office.

NCHS. M.D. Carroll, S. Abraham, and C.M. Dresser. 1983. *Dietary Intake Source Data, 1976–80.* Series II, no. 231. DHHS pub. no. (PHS) 83–1681. Washington, D.C.: U.S. Gov't Ptg. Office.

NOSEDA, A., M. ADLER, P. KETELBANT, D. BARAN, and J. WINAND. 1985. Massive vitamin A intoxication with ascites and pleural effusion. *J. Clin. Gastroenterol.* 7: 344–49.

NUTRITION REVIEWS. 1961. Editorial. Toxicity of vitamin K substitutes in premature infants. *Nutr. Rev.* 19: 75.

NUTRITION REVIEWS. 1983. Editorial. Megavitamin E supplementation and vitamin K–dependent carboxylation. *Nutr. Rev.* 41: 268–70.

NUTRITION REVIEWS. 1985. Editorial. Megavitamins and the hyperactive child. *Nutr. Rev.* 43: 105–7.

O'REILLY, R. and D. RYTAND. 1981. Resistance to warfarin due to unrecognized vitamin K supplementation. *New Engl. J. Med.* 303: 160–61.

ORWOLL, E.S. 1982. The milk-alkali syndrome; current concepts. *Ann. Intern. Med.* 97: 242–48.

OWEN, C.A. Jr., G.M. TYCE, E.V. FLOCK, and J.T. McCALL. 1970. Heparin–ascorbic acid antagonism. *Mayo Clin. Proc.* 45: 140.

PARFITT, A.M., J.C. GALLAGHER, R.P. HEANEY, C.C. JOHNSTON, R. NEER, and G.D. WHEDEN. 1982. Vitamin D and bone health in the elderly. *A.J.C.N.* 36: 1044–51.

PARRY, G.J. and D.E. BREDESEN. 1985. Sensory neuropathy with low-dose pyridoxine. *Neurology* 35: 1466–68.

PAULING, L. (Affirmative), C. MOERTEL (Negative). 1986. A proposition: Megadoses of vitamin C are valuable in the treatment of cancer. *Nutr. Rev.* 44: 28–31.

PAULING, L., A.B. ROBINSON, S.S. OXLEY, M.K. BERGESON, A. HARRIS, P. CARY, J. BLETHEN, and I.T. KEAVENY. 1973. Results of a loading test of ascorbic acid, niacinamide, and pyridoxine in schizo subjects and controls. In *Orthomolecular Psychiatry Treatment of Schizophrenia.* ed. D. Hawkins and L. Pauling. 18–34. San Francisco: W.H. Freeman.

PECK, G.L., T.G. OLSEN, F.W. YODER, J.S. STRAUSS, D.T. DOWNING, M. PANDYA,
D. BUTKUS, and J. ARNAUD-BATTANDIER. 1979. Prolonged remission of cystic
and conglobate acne with 13-cis-retinoic acid. *New Engl. J. Med.* 300: 329–33.

PHILLIPS, W.E., J.H.L. MILLS, S.M. CHARBOUNEAU, L. TRYPHONAS, G.V. HA-
TINA, Z. ZAWIDZHA, F.R. BRYCE, and I.C. MUNRO. 1978. Subacute toxicity of
pyridoxine hydrochloride in the beagle dog. *Tox. Appl. Pharmacol.* 44: 323–33.

RANDALL, R.E., M.D. COHEN, C.C. SPRAY, and E.C. ROSSMEISSL. 1964. Hypermagnesemia in renal failure. Etiology and toxic manifestations. *Ann. Int. Med.* 61:
73–88.

RANDLE, H.W., J.L. DIZA-PEREZ, and R.K. WINKELMANN. 1980. Toxic doses of
vitamin A for pityriasis rubrapilaris. *Arch. Dermatol.* 116: 888–92.

READ, M.H. and A.S. GRANEY. 1982. Food supplement usage by the elderly. *J.A.D.A.*
80: 250–53.

REPORT OF THE AD HOC COMMITTEE, National Nutrition Consortium, 1978.

REYNOLDS, E.H. 1973. Anticonvulsants, folic acid and epilepsy. *Lancet* 1: 1376.

REYNOLDS, E.H., R.H. MATTSON, and B.B. GALLAGHER. 1972. Relationship between serum and cerebrospinal fluid, anticonvulsant drug and folic acid concentrations
in epileptic patients. *Neurology* 22: 841–44.

RIMLAND, B., E. CALLAWAY, and P. DREYFUS. 1978. The effect of high doses of
vitamin B_6 on autistic children: a double blind cross-over study. *Am. J. Psychiat.* 135:
472–75.

ROBINSON, D.S. 1975. *Practical Drug Therapy,* 1st ed. 60–61. Burlington, VT: George
Little Press.

ROE, D.A. 1966. Nutrient toxicity with excessive intake. 1. Vitamins. *New York State
J. Med.* 66: 869–73.

_____. 1979. *Alcohol and the Diet.* 128–30. Westport, CT: AVI.

ROSENTHAL, G. 1971. Interaction of ascorbic acid and warfarin. *J.A.M.A.* 215: 1671.

ROWE, P.B. 1978. Inherited disorders of folate metabolism. In *The Metabolic Basis of
Inherited Disease,* 4th ed. ed. J.B. Stanbury, J.B. Wyngaarden, and D.S. Fredrickson.
pp. 430–58. New York: McGraw-Hill.

RUBY, L.K. and M.A. MITAL. 1974. Skeletal deformities following chronic hypervitaminosis A. *J. Bone and Joint Surg.* 56A: 1283–87.

RUDMAN, D. and P.J. WILLIAMS. 1983. Megadose vitamins. Use and misuse. *New
Engl. J. Med.* 309: 488–89.

RUFFIN, J.M. and D.T. SMITH. 1939. Treatment of pellagra with special reference to
nicotinic acid. *South. Med. J.* 32: 40.

SCHAUMBURG, H., J. KAPLAN, A. WINDEBANK, N. VICK, S. RASMUS, D. PLEA-
SURE, and M.J. BROWN. 1983. Sensory neuropathy from pyridoxine abuse. A new
megavitamin syndrome. *New Engl. J. Med.* 309: 445–48.

SMITH, I.J., M.F.G. BUCHANAN, I. GOSS, and P.J. CONGDON. 1983. Vitamin E in
retrolental fibroplasia. *New Engl. J. Med.* 309: 669.

SMITHELLS, R.W., S. SHEPPARD, C.J. SCHORAH, M.J. SELLER, N.C. NEIRN, R. HAMS, A.P. READ, and D.W. FIELDING. 1980. Possible prevention of neural tube defects by periconceptional vitamin supplementation. *Lancet* 1: 339–40.

SNEATH, P., I. CHANARIN, H.M. HODKINSON, C.K. McPHERSON, and E.H. REYNOLDS. 1973. Folate status in a geriatric population and its relation to dementia. *Age and Ageing* 2: 177–81.

STANTON, J.L. 1983. Vitamin usage: rampant or reasonable? Vitamin Issues: VNIS. Nutley, NJ: Hoffman LaRoche.

STEINKAMP, R.C., N.L. COHEN, and H.E. WALSH. 1965. Resurvey of an aging population—14 years follow-up. The San Mateo Nutrition Study. *J.A.D.A.* 46: 103–10.

SUGERMAN, A.A. and C.G. CLARK. 1974. Jaundice following administration of niacin. *J.A.M.A.* 228: 202.

TAKENOUCHI, K., K. ASO, K. KAWASE, H. IDRIKAWA, and T. SHIOMI. 1966. On the metabolites of ascorbic acid, especially oxalic acid, eliminated in urine, following the administration of large amounts of ascorbic acid. *J. Vitaminol. Japan* 12: 49.

TAUSSIG, H.B. 1966. Possible injury to the cardiovascular system from vitamin D. *Ann. Intern. Med.* 65: 1195.

U.S. DEPARTMENT OF AGRICULTURE. 1977–78. Nationwide Food Consumption Survey.

VAHLQUIST, A., E.B. BERU, and C. BERU. 1982. Skin content and plasma transport of vitamin A and beta-carotene in chronic renal failure. *Europ. J. Clin. Invest.* 12: 63–67.

VALENTIE, J.P., A.N. ELIAS, and G.D. WEINSTEIN. 1983. Hypercalcemia associated with oral isotretinoin in the treatment of severe acne. *J.A.M.A.* 250: 1899–1900.

VERNER, J.V. Jr., F.L. ENGEL, and H.J. McPHERSON. 1958. Vitamin D intoxication: Report of 2 cases treated with cortisone. *Ann. Int. Med.* 48: 765–73.

WALLACE, W.E., E. WILLOUGHBY, and P. BAKER. 1978. Coma in the Wernicke-Korsakafj syndrome. *Lancet* 2: 400–01.

WALSER, M. 1980. Calcium carbonate–induced effects on serum Ca × P product and serum creatinine in renal failure: A retrospective study. In *Phosphate and Minerals in Health and Disease,* ed. S.G. Massry, E. Ritz, and H. John. pp. 281–85; 675. New York: Plenum Press.

WEBER, F.L., G.E. MITCHELL, D.E. POWELL, B.J. REISER, and J.G. BANWELL. 1982. Reversible hepatotoxicity vitamin A accumulation in a protein deficient patient. *Gastroenterology.* 82: 118–123.

WHARTON, B.A. and S.J. DARKE. 1982. Infantile hypercalcemia. In *Adverse Effects of Foods,* ed. E.F.P. Jelliffe and D.B. Jelliffe. pp. 397–404. New York: Plenum Press.

WILKIN, J.K., O. WILKIN, R. KAPP, R. DONACHIE, M.E. CHENOWSKY, and J.

BUCKNER. 1981. Aspirin blocks nicotinic acid–induced flushing. *Clin. Pharmacol. Therap.* 31: 478–82.

WINDHORST, D.B. and T. NIGRA. 1982. General clinical toxicology of oral retinoids. *J. Am. Acad. Dermatol.* 6: 675–82.

WOLF, G. 1982. Is dietary β-carotene an anti-cancer agent? *Nutr. Rev.* 82: 118–23.

WOLSKA, H., S. JABLONSKA and Y. BOUNAMEAUX. 1983. Etretinate in severe psoriasis. *J. Am. Acad. Dermatol.* 9: 883–89.

Chapter 8

DRUG INTERFERENCE WITH THE BIOCHEMICAL ASSESSMENT OF NUTRITIONAL STATUS: ANALYTICAL AND BIOLOGICAL EFFECTS

Drugs can interfere with the biochemical assessment of nutritional status through analytical or biological effects (Roe 1981). Analytical effects include all interferences that result from chemical interactions in the test procedure (see Table 8.1). It is generally accepted that these effects are not caused by any in vivo nutritional changes caused by the drug. Conversely, biological interference involves a drug–nutrient interaction in vivo that affects nutritional status (see Table 8.2). Analytical effects involve a change in test values that is similar in all samples from patients who have received similar doses of the drug.

ANALYTICAL EFFECTS

International guidelines have been proposed for the evaluation of drug effects in clinical chemistry (Galteau and Siest 1984). In these guidelines, it is recommended that certain groups of drugs should be tested to determine whether they cause significant analytical interference with laboratory tests. According to this document, the drugs to be tested should include those most frequently used, those most frequently prescribed for the diseases monitored by the laboratory test under study, and those drugs that by their chemical nature are most likely to interfere with the test under investigation. While the guidelines are general, the recommendations are applicable to biochemical tests used specifically in the assessment of nutritional status. Drug interference is considered to be significant when the error induced by the drug is greater than the intra-assay imprecision (1 SD) of the method measured at the same level. Analytical interferences by drugs may be classified into several conditions as follows:

1   The drug reacts with a reagent to produce a chromophore. Examples include effects of cephalosporin antibiotics that interfere with creatinine determination.
2   The property is similar to that of the substance being measured. Examples include effects of ascorbic acid that, because it is a reducing agent, may inhibit the color reaction in glucose oxidase tests.
3   The drug absorbs at the same wavelength as the test substance. Examples include interferences by the folate antagonist methotrexate in methods that use an absorbance range of 340–410 nm.
4   The drug interferes with microbiological assays. Examples include antibiotics that interfere with the L. casei assay for plasma or erythrocyte folate.
5   The drug, as well as the nutrient, being assayed is fluorescent. An example is provided by tetracycline, which may interfere with the determination of urinary riboflavin.
6   The drug affects an enzyme activity. Oral contraceptives may affect erythrocyte aspartate aminotransferase activity, which is used as an enzyme assay for vitamin B_6, but the drug effect may occur without vitamin B_6 depletion (Rose et al. 1973; Letellier and Desjarlais 1985a).

There are numerous reports of drug interferences with laboratory tests, including test procedures for the assessment of nutritional status. The magnitude of the difference in test results is often small, however, and may vary with the instrument used for the analysis (Letellier and Desjarlais 1985b; Baer et al. 1983). Important analytical interferences are shown in Table 8.1.

BIOLOGICAL EFFECTS

DRUG-INDUCED STEATORRHEA

Drugs that have been shown to increase fecal fat excretion include alcohol, cholestyramine, neomycin, kanamycin, lincomycin, para-aminosalicylic acid, colchicine, phenolphthalein, bisacodyl, and methotrexate.

Cholestyramine, a basic anion-exchange resin used to treat hypercholesterolemia, forms nonabsorbable complexes with bile salts that are then lost in the feces. Fat absorption is reduced because of a lowered supply of bile salts for micelle formation and because the activity of pancreatic lipase is lessened when bile salts are unavailable, causing fat maldigestion. Cholestyramine-induced steatorrhea is quantita-

TABLE 8.1 ANALYTICAL INTERFERENCE WITH TESTS USED IN THE
ASSESSMENT OF NUTRITIONAL STATUS

Drug	Effect	Reference
Acetylsalicylic acid	Cholesterol, triglyceride, and iron increased in the plasma	Jelic-Ivanovic et al. 1985
Ibuprofen	Albumin, uric acid, and creatinine are decreased in plasma	Jelic-Ivanovic et al. 1985
Diclofenac, Indomethacin	Urea nitrogen increased	Jelic-Ivanovic et al. 1985
Cefoxitin, Cephalothin, Cephaloridine	Interference with the Jaffe reaction for creatinine	Letellier & Desjarlais 1985a
Calcium gluconate	Decreased plasma magnesium	Wacker & Parisi 1968
Ascorbic acid	Increased plasma uric acid	Caraway 1963
Canthaxanthine	Interference with the estimation of beta-carotene	Rock et al. 1981
Tetracycline	Interference with microbiological folate assay	Chanarin 1979
Methotrexate	Interference with radiometric folate assay	Waxman 1979

tively dose-dependent and is of clinical importance when very large doses of cholestyramine are administered (Hashim et al. 1961; West and Lloyd 1975). Colestipol, another bile acid sequestrant, has not been shown to induce significant steatorrhea with conventional dosage schedules (Glueck et al. 1972).

Neomycin causes steatorrhea because of intestinal mucosal damage, and because it precipitates the sodium salts of taurocholic as well as glycoholic and desoxycholic acids, rendering bile salts less available for fat absorption. Polymyxin, kanamycin, and lincomycin can also induce steatorrhea, probably because of mucosal damage (Bartelink 1974).

Para-aminosalicylic acid (PAS) produces steatorrhea, possibly also because of bile acid chelation (Levine 1968). Other drugs causing villous damage and impaired regeneration of the epithelial cells of the small intestine cause steatorrhea, including colchicine, methotrexate, and other cancer-chemotherapeutic agents (Roe 1976).

Cathartics or laxatives used to excess may cause steatorrhea, most commonly phenolphthalein and bisacodyl (Heizer et al. 1968).

TABLE 8.2 DRUGS THAT ELEVATE, DEPRESS, OR INTERFERE WITH
MEASUREMENT OF BLOOD LEVELS OF VITAMINS

Drug	Vitamin Level Affected
Mineral oil	Plasma carotene
Canthaxanthine	Plasma carotene, plasma retinol
Cholestyramine, Colestipol	Plasma retinol
Oral contraceptives	Plasma retinol
Phenytoin, Phenobarbital, Corticosteroids	Plasma 25-hydroxycholecalciferol
Aspirin, Oral contraceptives	Plasma and leukocyte ascorbate
Neomycin, Cholestyramine, Potassium chloride, Phenformin, Metformin, p-Aminosalicylic acid, Colchicine, Cimetidine, Broad-spectrum antibiotics	Plasma/serum vitamin B_{12}
Ethanol, Aspirin	Plasma folacin
Phenytoin Phenobarbital, Glutethimide, Cholestyramine, Colestipol, p-Aminosalicylic acid, Sulfasalazine, Oral contraceptives, Methotrexate	Plasma and erythrocyte folate

*DRUGS AFFECTING PLASMA, ERYTHROCYTE, AND
LEUKOCYTE LEVELS OF VITAMINS*

Vitamin A and Carotene

Curtis and Ballmer (1939) demonstrated that serum carotene levels
from diet are reduced when either mineral oil or a mineral oil emulsion
(20 ml) is given before meals.

A recent report has shown that carotenoid, used as a tanning agent,
not only colors the plasma orange, but also interferes with measurement of serum levels of carotene and vitamin A (see Table 8.2). Canthaxanthine, a carotenoid, is used as a food colorant. Capsules containing canthaxanthine are also being used to produce pseudotanning
of the skin. After discontinuance of the capsule, the canthaxanthine

may remain in the plasma for about ten days, making determination of beta-carotene and vitamin A difficult during this period, as well as during the period of intake (Rock et al. 1981).

Drugs causing malabsorption may lower plasma retinol levels. This effect has been documented with respect to neomycin, cholestyramine, and colestipol (Longnecker and Basu 1965; Schwarz et al. 1980). Serum retinol levels are lowered with prolonged treatment with these bile-acid sequestrants, but values remain within normal limits.

Oral contraceptives consistently elevate plasma retinol levels, and this change is associated with increased levels of plasma retinol–binding protein (Gal et al. 1971).

Vitamin E

Plasma tocopherol levels may be reduced by hypolipidemic drugs. When cholestyramine is given to children with familial hypercholesterolemia, mean serum tocopherol levels decrease over the first one to two years of therapy, but values still remain within the normal range (West and Lloyd 1975).

Vitamin D

Several drugs have been shown to lower serum 25-hydroxycholecalciferol levels. There have been reports of children and adults with seizure disorders who showed reduced levels of 25-hydroxycholecalciferol when they were taking phenytoin and phenobarbital. Plasma 25-hydroxycholecalciferol levels may be lowered in children receiving corticosteroids (Chesney et al. 1978; Hahn et al. 1972).

Isoniazid can be considered as a vitamin D antagonist. This drug depresses serum levels of vitamin D metabolites and imposes a risk of osteomalacia with prolonged use (Bengoa et al. 1984).

Vitamin C

Plasma ascorbate levels may be reduced in children and adults taking large quantities of aspirin because aspirin potentiates the excretion of vitamin C (Coffey and Wilson 1975; Daniels and Everson 1936–37). It has been reported that plasma and leukocyte ascorbate levels are lowered by oral contraceptives, but these reports were published when high-estrogen oral contraceptives were being used. Whether or not low-estrogen oral contraceptives cause a significant reduction in plasma and leukocyte ascorbate levels is unclear (Briggs and Briggs 1972; Rivers and Devine 1972; Roe 1977).

Vitamin B_{12}

Plasma or serum levels of vitamin B_{12} are reduced by drugs that reduce the digestibility of foods containing vitamin B_{12} or impair the vitamin's absorption. Most drugs that cause vitamin B_{12} malabsorption have their effect in the terminal ileum. Ileal malabsorption of vitamin B_{12} has been reported following administration of neomycin, cholestyramine, potassium chloride, phenformin and metformin, para-aminosalicylic acid, and colchicine (Roe 1979).

Prolonged cimetidine therapy suppresses gastric acidity and production of pepsin, thus impairing release of vitamin B_{12} from animal protein foods and consequently lowering serum vitamin B_{12} levels (McGuigan 1980). Bacterial assays (L. leichmanii) for vitamin B_{12} are sensitive to antibiotics and cytotoxic agents. However, it has been stated that either protozoal assays (*Euglena gracilis* or *Ochromonas malhamensis*) are not affected by these drugs, or drugs (cytotoxic agents) that might affect the assays are "diluted out" (Baker et al. 1981).

Folacin (Folic Acid)

Drug-induced folacin deficiency may be mild and without evidence of megaloblastic anemia. Severe chronic drug-induced folacin deficiency is associated with diminished levels of plasma and erythrocyte folacin, but not all drug-related megaloblastic anemias are due to folacin deficiency (Stebbins et al. 1973). However, the drug most commonly responsible for acute lowering of plasma folacin levels is ethanol, sometimes when used as a parenteral energy source (Eichner and Hillman 1973).

Plasma folacin may be lowered by aspirin ingestion (Alter et al. 1971; Lawrence et al. 1984). Plasma and erythrocyte folacin levels may be lowered in patients receiving anticonvulsant drugs, including phenytoin and phenobarbital as well as glutethimide (Reynolds et al. 1966). Folacin deficiency associated with intake of phenytoin is due to enhanced catabolism of folacin (Krumdieck 1978). Malabsorption of folacin causing low or deficient plasma and erythrocyte folacin levels may occur with prolonged intakes of bile-acid sequestrants, antacids, para-aminosalicylic acid, and sulfasalazine (West and Lloyd 1975; Roe 1976; Franklin and Rosenberg 1973). Plasma and erythrocyte folacin levels have frequently been reported to be lower in women on oral contraceptives compared with women not receiving these drugs. The magnitude of the effect of oral contraceptives has diminished with use of the low-estrogen preparation (Roe 1977; Shojania et al. 1971). Folacin antagonists cause a reduction in plasma and erythrocyte folacin levels with chronic administration.

Spurious "abnormal" folacin values may be reported when radiometric assays for folacin are employed to measure blood folacin levels in patients on a drug such as methotrexate, because of altered folacin affinity for the protein binder (Waxman 1979). "Abnormal" folacin values, caused by assay interference, also may be reported when the microbiologic assay (*L. casei*) is used for the estimations when patients are receiving antibiotics, particularly broad-spectrum antibiotics (Chanarin 1979).

DRUGS AND RED CELL ENZYME TESTS OF VITAMIN STATUS

Red cell enzyme tests are commonly used to assess thiamin, riboflavin, and vitamin B_6 nutriture. The respective red cell enzyme systems are erythrocyte transketolase for thiamin, glutathione reductase for riboflavin, and either aspartate aminotransferase or alanine aminotransferase for vitamin B_6. These enzymes require their respective coenzymes, such as thiamin pyrophosphate, flavin adenine dinucleotide, and pyridoxal phosphate, for activity. In vitro, red cell enzyme tests involving these coenzymes are used to estimate the extent of unsaturation of the in vivo enzyme systems. In all these tests of vitamin status, enzyme activities are measured in vitro, both before and after addition of the respective coenzyme (Thurnham 1981). In the erythrocyte pyridoxamine oxidase assay for riboflavin status, the enzyme, riboflavin-5'-phosphate, is added in vitro to obtain the activity coefficient (Clements and Anderson 1980).

It has been claimed, with reasonable justification, that these red cell enzyme assay tests for B vitamin nutriture are the most sensitive systems to be affected by early deficiency of the vitamins. Red cell enzyme activities decline as the red cells age, and the red cell activity coefficients are elevated with red cell age. It has been proposed that interday and interpersonal variability in red cell enzyme activity may be attributed to change in the age structure of the red cell population (Spooner et al. 1979).

Drugs may affect red cell enzyme activity because (1) they impair vitamin status, (2) they alter apoenzyme activity, (3) they reduce the life span of the red cell, or (4) they actually interfere with the assay procedure.

Biochemical evidence of vitamin B_6 depletion has been reported in women receiving oral contraceptives (Price et al. 1967; Rose 1966).

Erythrocyte alanine aminotransferase activity and the stimulation of this enzyme activity in vitro by pyridoxal phosphate has been compared in women taking an estrogen-containing oral contraceptive and in women not on the drug. No differences in basal enzyme activity were shown between the two groups, but those using oral contraceptives had had a significantly higher percent stimulation in vitro. This was considered by the investigators to mean that a subclinical deficiency of vitamin B_6 was present. In this same study, erythrocyte aspartate aminotransferase activity was elevated in the group on oral contraceptives, but the in vitro stimulation was unchanged. Treatment of women on oral contraceptives with pyridoxine hydrochloride (40 mg/day for four to eight weeks) produced elevations in the activity of both of these transaminases and decreased in vitro stimulation. It was proposed that these enzyme changes resulted both from activation of latent apoenzyme and cofactor stimulation (Rose et al. 1973).

The metabolism of oral L-tryptophan and L-kynurenine has been studied in women on oral contraceptives. Both caused higher urinary excretion of several metabolites of the tryptophan–nicotinic acid ribonucleotide pathway in the contraceptive users compared with controls. Two mechanisms were previously proposed to account for these changes, including steroid-induced reduction in the activity of tryptophan oxygenase and interference with the coenzyme function of pyridoxal phosphate at the kynureninase-catalyzed step of the pathway, which is necessary for the conversion of 3-hydroxykynurenine to 3-hy-droxyanthranilic acid (Leklem et al. 1973). The authors suggest that contraceptive steroids increase the rate of renal loss and excretion of kynurenine and 3-hydroxykynurenine so that these tryptophan metabolites are not available for transamination.

Estrogen may inhibit kynureninase, causing hyperexcretion of metabolites proximal to this step in the pathway (*Nutr. Rev.* 1973). More recently, inhibition of kynureninase by estrone sulfate has been demonstrated, and this is now being considered as a possible explanation for abnormal tryptophan and kynurenine load tests in women on oral contraceptives (Bender and Wynick 1981).

Other drugs that can alter red cell enzyme tests are 5-fluorouracil and furosemide, both of which affect the erythrocyte transketolase assay (Basu et al. 1979; Yui 1980).

During alcohol intoxication, thiamin levels in whole blood decrease and the thiamin diphosphate effect on transketolase activity increases. The mean thiamin diphosphate effect has been shown to be significantly elevated in alcoholic patients imbibing alcohol; yet, the erythrocyte transketolase activity in alcoholic patients is low and does not

TABLE 8.3 DRUGS THAT ARE VITAMIN ANTAGONISTS

Folate (folacin) antagonists	Methotrexate
	Pyrimethamine
	Triamterene
	Trimethoprim
	Pentamidine isethionate
	Sulfasalazine
	Triazinate
Vitamin B_6 antagonists	Isoniazid
	Hydralazine
	Cycloserine
	Levodopa
	Penicillamine
Riboflavin antagonist	Boric acid
Vitamin B_{12} antagonist	Nitrous oxide
Vitamin K antagonists	Coumarin anticoagulants
	Cephalosporin antibiotics
Vitamin D antagonists	Isoniazid
	Phenytoin
	Phenobarbital

Source: Roe 1979.

normalize rapidly with administration of thiamin (Waldenlend et al. 1980).

VITAMIN ANTAGONISTS AFFECTING TESTS OF VITAMIN STATUS

The antivitamin effects of drugs, sometimes used intentionally in the treatment of disease, may have unwanted side effects (Roe 1979).

Drugs common for their effects as vitamin antagonists are methotrexate, used in the treatment of choriocarcinoma, head and neck cancer, and acute lymphoblastic leukemia; and pyrimethamine, used in the prevention of malaria. Others are listed in Table 8.3.

Methotrexate has multiple antifolate effects. In the tissues, the drug binds tightly to the dihydrofolate reductase enzyme. Folate is displaced from the enzyme by the drug and is excreted in the urine. Methotrexate polyglutamates are formed and synthesis of folate polyglutamates is impaired. Thymidylate synthetase is inhibited. DNA, RNA, and protein syntheses are inhibited. Methotrexate reduces the incorporation of deoxyuridine (dU) into DNA and favors incorporation of [^{3}H]thymidine into DNA by the alternate pathway. Therefore, the dU suppression test is abnormal in patients receiving methotrexate (Wickramasinghe and Saunders 1977).

Moreover, methotrexate has a greater affinity for folate-binding protein at the pH optimum for the radiometric assay for serum (plasma) folacin. Therefore, it has been proposed that the radiometric method, which is a competitive protein-binding assay in plasma and erythrocytes, should not be used to determine folacin in plasma and erythrocytes in patients on this drug (Thurnham 1981).

Sulfasalazine is a folate antagonist. Sulfasalazine inhibits the intestinal absorption of dietary folacin by competing for the folacin transport system, causing patients receiving the drug to develop folate deficiency (Franklin and Rosenberg 1973). Sulfasalazine also inhibits three enzymes in vitro, including dihydrofolate reductase, methyltetrahydrofolate reductase, and serine transhydroxymethylase, each of which catalyze a different reaction involving folate coenzymes.

Sulfasalazine has been shown to act as a folate antagonist in intact lymphocytes. The drug inhibits serine synthesis in rat spleen lymphocytes, apparently as a result of inhibition of folate-dependent pathways. The drug also interferes with the dU suppression test. Other drugs that inhibit the dihydrofolate reductase enzyme do not produce an abnormal dU suppression test when the serum vitamin B_{12} levels and erythrocyte folacin levels are normal (Baum et al. 1981; Berglund and Andersson 1981).

Vitamin K antagonists include not only coumarin anticoagulants that are intentionally used as antivitamins, but also certain antibiotics where the anticoagulant effect is an adverse drug reaction. Cephalosporin antibiotics are also vitamin K antagonists and, as such, can cause hemorrhagic episodes (Roe 1984).

DRUG-RELATED CHANGES IN ELECTROLYTE BALANCE AND MINERAL STATUS

Serum Sodium Levels (See Table 8.4)

Hyponatremia.   Common mechanisms responsible for hyponatremia include (1) an increase in body water (dilutional hyponatremia); (2) a decrease in body solute (sodium net loss); and (3) addition of solute to plasma (serum), causing an osmotic redistribution of water (Flear et al. 1981). Drugs can cause all three types of hyponatremia. Drug-induced dilutional hyponatremia is caused by inappropriate secretion of antidiuretic hormone (SIADH) (Moses and Miller 1974).

The antidiuretic action of oral hypoglycemic agents has frequently been reported (Hagan and Frawley 1970). Chlorpropamide has been shown to have an antidiuretic effect in patients with both diabetes insipidus and diabetes mellitus. With this drug the hyponatremia simulates effects of ADH in laboratory rats. Tolbutamide can produce a

TABLE 8.4 DRUGS THAT INDUCE CHANGES IN
SERUM SODIUM LEVELS

Drugs that increase sodium level (Hypernatremia)	Drugs that decrease sodium level (Hyponatremia)
Diazoxide	Chlorpropamide
Carbenicillin	Tolbutamide
Sodium sulfate	Vincristine
	Cyclophosphamide
	Carbamazepine
	Amitriptyline
	Thioridazine
	Clofibrate
	Hydrochlorothiazide
	Polythiazide and other thiazides
	Metalazone
	Mannitol
	Spironolactone
	Captopril
	Licorice

similar type of dilutional hyponatremia (Luethi and Studer 1969). The biguanides phenformin and metformin may show antidiuretic effects in patients with diabetes insipidus, apparently by potentiating the effect of ADH.

Two cancer chemotherapeutic agents have been shown to produce SIADH. These are vincristine, which usually causes water retention within one to two weeks following the initiation of treatment (Oldham and Pomeroy 1972), and cyclophosphamide, which can produce its antidiuretic effect beginning 4 to 12 hours after injection and lasting for up to 20 hours (Steele et al. 1973).

Tricyclic compounds that produce SIADH include carbamazepine, used as an anticonvulsant (Rado 1973), as well as amitryptyline, used as an antidepressant (Luzecky et al. 1974).

The hypolipidemic agent clofibrate exerts an antidiuretic effect in patients with diabetes insipidus and also may inhibit water excretion after loading in normal people (Moses et al. 1973).

Hyponatremia from bodily sodium depletion can occur with diuretic therapy. Thiazides have been reported as the cause of serious and fatal hyponatremia in elderly patients. Hyponatremia with these drugs develops slowly and will usually resolve rapidly with discontinuance of the drug unless the patient is drinking large volumes of water (Ashraf et al. 1981). In patients with diabetes insipidus, thiazide diuretic may exhibit an antidiuretic action (Crawford and Kennedy 1959).

Spironolactone can cause hyponatremia with an associated rise in serum potassium levels. Captopril may also cause hyponatremia; both drugs may have an anti-aldosterone effect (Nicholls et al. 1980).

Mannitol infusion can cause hyponatremia because the infused molecules remain within the vascular compartment and raise plasma osmolality. There is an expansion of blood volume and an associated lowering of sodium concentration.

Hypernatremia.   Hypernatremia can be induced by diazoxide, a thiazide derivative, which has been used in the treatment of hypertension. Diazoxide may increase the proximal tubular reabsorption of sodium as well as water (Bartorelli et al. 1963).

The semisynthetic penicillin carbenicillin has been reported to cause hypernatremia as a toxic side effect (Brumfit and Percival 1967).

Sodium sulfate infusion, used in the treatment of hypercalcemia, can cause both hypernatremia and hypokalemia. Risks are greatest in patients with cardiovascular disease, and close monitoring of serum electrolytes is essential in these patients if sodium sulfate is administered (Sherwood 1967).

Serum Potassium Levels (See Table 8.5)

Hypokalemia.   Potassium depletion has long been recognized as a potentially serious side effect of oral diuretic treatment (Hamdy et al. 1980). Diuretic drugs causing hypokalemia include thiazides and the loop diuretics, furosemide and ethacrynic acid. Hypokalemia is a particular risk in elderly patients receiving these diuretics because of low potassium intake and reduced muscle mass, which reduces total-body potassium reserves (Ramsey et al. 1977).

A British report on the incidence of hypokalemia secondary to oral diuretics has concluded that severe hypokalemia is uncommon and that the frequency of hypokalemia with diuretics depends on the initial serum potassium level as well as on its decrease (Morgan and Davidson 1980). The authors assert that the diagnosis of diuretic-induced hypokalemia would almost disappear if the serum potassium level indicating deficiency on laboratory reports were reduced from <3.5 mmol/L (13.7 mg/dl) to <3.0 mmol/L (11.7 mg/dl). Whether or not intake of thiazides or loop diuretics leads to severe hypokalemia is related to concurrent alcohol and laxative abuse, as well as to the presence of catabolic states which potentiate loss of potassium (Nardone et al. 1978).

Long-term diuretic therapy can cause a failure of renal conservation

TABLE 8.5 DRUGS THAT INDUCE CHANGES IN
SERUM POTASSIUM LEVELS

Drugs that increase potassium level (Hyperkalemia)	Drugs that decrease potassium level (Hypokalemia)
Hypertonic mannitol	Amphotericin B
Aminocaproic acid	p-Aminosalicylic acid
Isoniazid	Carbenoxolone
Phenformin	Gentamicin
Spironolactone	Lithium carbonate
Succinylcholine	Chlorthalidone
Triamterene	Furosemide
Indomethacin	Ethacrynic acid
Amiloride	Hydrochlorothiazide and other thiazides
	Albuterol (Salbutamol)
	Corticosteroids
	Salicylates (aspirin)
	Tetracycline (degraded)

of potassium and magnesium with a simultaneous loss of these ions from the tissues (Dyckner and Wester 1985).

Several reports have indicated that glucose intolerance developing in hypertensive patients on long-term thiazide therapy is secondary to potassium deficiency with hypokalemia (Amery et al. 1978). However, Swedish investigators Berglund and Andersson (1981) who carried out a 6-year follow-up of 53 hypertensive patients receiving either a thiazide drug or a beta-blocking agent, found no change in serum potassium during the first year of thiazide therapy. After 6 years of treatment 10% of these patients had hypokalemia (defined by this laboratory as serum potassium <3.6 mmol/L or 14.1 mg/dl). Propranolol, the beta-blocking agent employed in this study, did not cause hypokalemia. During the study, diabetes developed in one patient receiving thiazide and in another receiving propranolol. Fasting blood sugar did not change during follow-up of either of the two groups, except in the two people who became diabetic.

This study was criticized by Greveson et al. (1981) and by Jopling (1981), who pointed out that in the Swedish study the dose of thiazide drug was low, that it contained a potassium supplement, and that the patients were males, who are less susceptible to hypokalemia.

Hypokalemia is associated with laxative abuse because of loss of potassium into the gastrointestinal tract and failure of potassium reabsorption from the colon. Patients on excessive intakes of phenol-phthalein, bisacodyl, and senna have been reported with severe hypokalemia for which the cause was undiscovered until their habit of laxative abuse was suspected and/or proven (Fleming et al. 1975; Levine et al. 1981). Drugs that induce malabsorption, such as para-aminosalicylic acid, tend to produce hypokalemia.

Hypokalemia may occur during prolonged, high-dosage corticosteroid therapy. Potassium is mobilized from the tissues and excreted in the urine (Thorn 1966). Hypokalemia, hypochloremic alkalosis may supervene with administration of corticosteroids. Hypokalemia related to the use of potassium-losing diuretics is potentiated by administration of corticoids (Srivastava et al. 1973).

Hypokalemia can also result from intake of large quantities of licorice or from use of related drugs, such as carbenoxolone. Carbenoxolone is a triterpine and the disodium salt of the succinic acid ester of enoxolone (18 beta-glycerrhetic acid). Both this drug and licorice root preparations have been used in the treatment of dyspepsia and peptic ulcer, and both have mineralocorticoid-like actions (Gross et al. 1966; Lewis 1974).

Hypokalemia may be an outcome of treatment with a number of antibiotics that exhibit nephrotoxic side effects. Included in this group of drugs are amphotericin B (Drutz et al. 1970), gentamicin, and degraded tetracycline, which has been shown to produce a Fanconi syndrome (Mavromatis 1965).

High-dose salicylate therapy or salicylate overdosage can produce hypokalemia both because of cellular damage and potassium leakage from cells and because of the direct effect of other drugs on renal tubules (Smith and Smith 1966).

O'Brien et al. (1981) described a patient, an elderly woman, who developed hypokalemia after she took an overdose of the beta-adrenergic antagonist, solbutamol. This drug, in common with catecholamines and other sympathomimetic agents, causes transfer of potassium from the extracellular to the intracellular compartment by activating sodium-potassium-dependent ATPase. Hypokalemia has previously been described when the drug has been given intravenously to insulin-dependent diabetics and nondiabetics (Gungdogdu et al. 1979).

Hyperkalemia. Administration of succinyl choline can produce significant increases in serum potassium levels, which can be sufficiently severe to cause arrhythmia and cardiac arrest in trauma and burn patients (Gronert 1970; Roth and Wuthrich 1969). Large doses of isoniazid also may produce hyperkalemia (Baum et al. 1981). Hyperkalemia with metabolic acidosis can occur when biguanides such as phen-

TABLE 8.6 DRUGS THAT INDUCE CHANGES IN
SERUM CALCIUM LEVEL

Drugs that increase calcium level (Hypercalcemia)	Drugs that decrease calcium level (Hypocalcemia)
Thiazides	Phosphate laxatives
Vitamin D and metabolites	Sodium phytate
Isotretinoin	Sodium cellulose phosphate
Etretinate	EDTA
Calcium carbonate	Furosemide
	Ethacrynic acid
	Phenobarbital
	Phenytoin
	Phenolphthalein
	Bisacodyl
	Primidone
	Glutethimide
	Corticosteroids

formin are given to diabetics with impaired renal function (Mestman et al. 1969).

Hypertonic infusions of mannitol increase serum potassium levels (Moreno et al. 1969). The hyperkalemia resulting from administration of spironolactone is caused by this aldosterone antagonist blocking distal tubular sodium–potassium exchange. Hyperkalemia due to spironolactone is more severe in patients with impaired renal function (Herman and Rado 1966). Triamterene, another potassium-sparing diuretic, can also cause hyperkalemia (Dorph and Olgaard 1968). Dose-dependent increases in serum potassium have been reported with the use of beta blockers (Pederson and Kornerup 1976; Pederson and Mikkelsen 1979).

Serum Calcium Levels (See Table 8.6)

Hypocalcemia.   Phosphate salts given either intravenously or by oral administration reduce levels of serum calcium. An oral laxative containing phosphate, used to prepare patients for barium enemas, has been shown to elevate serum phosphorus levels; a decrease in serum calcium follows (Goldsmith and Ingbar 1966; Herbert et al. 1966; Wilberg et al. 1978). Nonnutrient dietary substances that have been used for their hypocalcemic effect include sodium phytate and sodium cellulose phosphate. EDTA was formerly used because of its chelating properties as a hypocalcemic agent in hypercalcemia (Milne 1974; Spencer et al. 1956).

Furosemide and ethacrynic acid decrease serum calcium levels (Eknoyan et al. 1970). Deliberate use of intravenous furosemide to treat hypercalcemia has resulted in a reduction in serum calcium in some patients, but complications of this treatment include hypomagnesemia and a decline in renal function (Suki et al. 1970).

Drugs that decrease calcium absorption and that may cause hypocalcemia include phenobarbital, phenytoin (including phenolphthalein and bisacodyl, which can cause steatorrhea in laxative abusers), primidone, glutethimide, and corticosteroids as well as drugs causing steatorrhea (Parke and Ioannides 1981; Roe 1974).

Hypercalcemia. Hypercalcemia may occur with administration of thiazide diuretics, but is relatively uncommon and may be a temporary effect (Duarte et al. 1971).

Hypercalcemia and bone demineralization occurs with excessive intake of vitamin D and its metabolites (De Luca 1980). Vitamin D–induced hypercalcemia is an important cause of polyuric renal failure. When high doses of vitamin D are taken, the hypercalcemic effect is potentiated by concurrent intake of calcium salts (Milne 1974).

Hypercalcemia has been reported in patients receiving the synthetic retinoids, isotretinoin and etretinate. The hypercalcemia may be associated with osteophyte formation and calcification of the spinal ligaments. Both syndromes have also been reported with prolonged intake of very high doses of vitamin A (Gerber et al. 1984; Gerber et al. 1954; Pittsley and Yoder 1983). Intake of large amounts of dietary calcium with an antacid, or of calcium carbonate, may induce the development of the milk-alkali syndrome, which is characterized by hypercalcemia and renal impairment (Orwoll 1982).

Malignant tumors that metastasize to bone may induce hypercalcemia. When androgenic steroids are used in the management of metastasis from breast tumors, hypercalcemia may be increased (Spencer and Lewin 1963).

Serum Magnesium Levels (See Table 8.7)

Hypomagnesemia. Drugs causing hypomagnesemia include diuretics, antibiotics, and chemotherapeutic agents (Flink 1985). Hypomagnesemia may be related to intake of thiazides or loop diuretics. Thiazide-induced hypomagnesemia occurs frequently when the daily dose is high and when magnesium intake from food is low (Duarte 1968). A common additional risk factor is alcohol abuse, since high alcohol intake also induces magnesium depletion (Sullivan et al. 1963). Hypomagnesemia associated with intake of thiazides and the loop diuretics, ethacrynic acid and furosemide, is secondary to urinary hy-

TABLE 8.7 DRUGS THAT INDUCE CHANGES IN
SERUM MAGNESIUM LEVELS

Drugs that increase magnesium level (Hypermagnesemia)	Drugs that decrease magnesium level (Hypomagnesemia)
Magnesium sulfate	Thiazides
Magnesium antacids	Furosemide
Lithium carbonate	Ethacrynic acid
	Ammonium chloride
	Mercurial diuretics
	Neomycin
	Gentamicin
	Cephalothin
	Methoxyflurane
	Ethanol
	CIS-platinum

perexcretion of magnesium. When hypomagnesemia is found in patients with congestive heart failure, who are on diuretics and digitalis glycosides, digitalis toxicity with arrhythmia may be secondary to the magnesium deficiency. Hypomagnesemia as well as a significant magnesium deficit in erythrocytes, skeletal muscle, and bone may develop in patients on long-term diuretic therapy for heart failure because of the combined effects of diuretics, low magnesium intake, digoxin therapy, and secondary aldosteronism (Wacker and Parisi 1968; Lim and Jacob 1972). Combined hypomagnesemia, hypokalemia, and hyponatremia are particularly common in patients on digitalis for congestive heart failure (Whang et al. 1985).

Drugs causing significant steatorrhea also tend to increase fecal magnesium loss, probably owing to formation of magnesium soaps (Hansten 1973). Hypomagnesemia in patients on the antibiotic neomycin is secondary to maldigestion; with other antibiotics, hypomagnesemia is caused by urinary hyperexcretion of magnesium owing to the nephrotoxicity of these antimicrobial agents (Appel and Neu 1977).

There have been several reports of combined hypomagnesemia, hypokalemia, hypocalcemia, and alkalosis in patients receiving gentamicin (Bar et al. 1975). Gentamicin-induced renal tubular injury causing these metabolic defects is potentiated by cephalothin and methoxyflurane (Luft et al. 1976; Mazze and Cousins 1973).

The cancer chemotherapeutic agent CIS-platinum (II) dichloridiamine (CPDD) is nephrotoxic. Its nephrotoxicity is similar to that induced by heavy metal poisoning. CPDD nephrotoxicity is associated with a decreased glomerular filtration rate and loss of renal tubular absorptive capacity. Hyperexcretion of magnesium and hypomagnesemia have been reported during CPDD therapy. The nephrotoxic effect of CPDD and the associated magnesium depletion are prevented largely by adequate hydration of patients prior to and during the treatment period (Schilsky and Anderson 1979; Stark and Howell 1978).

Hypermagnesemia.   Hypermagnesemia can develop in patients with renal insufficiency who are given large doses of antacids containing magnesium sulfate as a cathartic (Alfrey et al. 1970; Ditzler 1970). Hypermagnesemia may develop during lithium carbonate therapy (Nielson 1964).

Serum Phosphorus Levels

Hypophosphatemia.   Hypophosphatemia has been induced by antacid abuse (Lotz et al. 1968). Phosphate depletion associated with excess intake of antacids would appear to be uncommon if the paucity of reports reflects the incidence. However, it is very possible that antacid-induced hypophosphatemia is either unrecognized or underreported. The antacids that can cause hypophosphatemia include aluminum and magnesium hydroxide. These hydroxides form nonabsorbable phosphates in the gut lumen. Secondary osteomalacia may develop (Insogna et al. 1980). Large doses of antacids have been used to treat hyperphosphatemia in patients with renal failure (Gabriel 1977).

Hyperphosphatemia.   Chronic analgesic nephropathy, owing to phenacetin, acetaminophen, or these drugs in combination with salicylates, is characterized by hyperphosphatemia when renal failure is imminent (Abel 1971; Koch-Weser 1976).

REFERENCES

ABEL, J.A. 1971. Analgesic nephropathy: A review of the literature. *Clin. Pharm. Therap.* 12: 583.

ALFREY, A.C., D.S. TERMAN, L. BRETTSCHNEIDER, K. SIMPSON, and D.A. OGDEN. 1970. Hypermagnesemia after renal homotransplantation. *Ann. Intern. Med.* 73: 367–71.

ALTER, H.J., M.J. ZVAIFER, and C.E. RATH. 1971. Interrelationship of rheumatoid arthritis, folic acid and aspirin. *Blood* 38: 405.

AMERY, A., C. BULPITT, A. DE SCHAEPDRYVER, R. FAGARD, J. HELLEMANS, A. MUTSERS, P. BERTHAUX, M. DERUYTTERE, C. DOLLERY, F. FORETTE, P. LUND-JOHANSEN, and J. TUOMILEHTO. 1978. Glucose tolerance during diuretic therapy. Results of trial by the European Working Party on Hypertension in the Elderly. *Lancet* 1: 681.

APPEL, G.B. and H.C. NEU. 1977. The nephrotoxicity of antimicrobial agents (2nd of 3 parts). *New Engl. J. Med.* 296: 721.

ASHRAF, N., R. LOCKSLEY, and A.I. ARIEFF. 1981. Thiazide-induced hyponatremia associated with death or neurological damage in outpatients. *Am. J. Med.* 70: 1163.

BAER, D., R.N. JONES, J.P. MULLOOLY, and W. HORNER. 1983. Protocol for the study of drug interferences in laboratory tests: Cefotaxime interference in 24 clinical tests. *Clin. Chem.* 29: 1736–40.

BAKER, H., O. FRANK, and S.H. HUTNER. 1981. Problems with the serum vitamin B_{12} assay. Letter. *Lancet* 1: 154.

BAR, R.S., H.E. WILSON, and E. L. MAZZAFERRI. 1975. Hypomagnesemic hypocalcemia secondary to renal magnesium wasting: A possible consequence of high-dose gentamycin therapy. *Ann. Intern. Med.* 82: 646.

BARTELINK, A. 1974. Clinical drug interactions in the gastrointestinal tract. In *Clinical Effects of Interaction Between Drugs.* ed. L.E. Cluff and J.C. Petrie. 103. Amsterdam: Excerpta Medica.

BARTORELLI, C., N. GARGANO, G. LEONETTI, and A. ZANCHETTI. 1963. Hypotensive and renal effects of diazoxide: A sodium-retaining benzothiadiazine compound. *Circulation* 27: 895–903.

BASU, T.K., M. AKSAY, and J.W. DICKERSON. 1979. Effects of 5-fluoracil on the thiamin status of adult female rats. *Chemotherapeutics* 25: 70.

BAUM, C.L., J. SELHUB, and I.H. ROSENBERG. 1981. Antifolate actions of sulfa-salazine on intact lymphocytes. *J. Lab. Clin. Med.* 97: 778.

BENDER, D.A. and D. WYNICK. 1981. Inhibition of kynureninase (L-kynurenine hydrolase EC3.71.3) by oestrone sulfate: An alternative explanation for abnormal results of tryptophan load tests in women receiving oestrogenic steroids. *Proc. Nutr. Soc.* 45: 269.

BENGOA, J.M., M.J.G. BOLT, and I.H. ROSENBERG. 1984. Hepatic vitamin D 25-hydroxylase inhibition by cimetidine and isoniazid. *J. Lab. Clin. Med.* 104:546–52.

BERGLUND, G. and O. ANDERSSON. 1981. Beta blockers or diuretics in hypertension? A six year follow up of blood pressure and metabolic side effects. *Lancet* 1: 744.

BRIGGS, M. and M. BRIGGS. 1972. Vitamin C requirements and oral contraceptives. *Nature* 238: 277.

BRUMFIT, W. and A. PERCIVAL. 1967. Clinical and laboratory studies with carbenicillin. *Lancet* 1: 1289.

CARAWAY, W.T. 1963. Uric acid. *Stand. Meth. Clin. Chem.* 4: 239.

CHARNARIN, I. 1979. *The Megaloblastic Anemias.* Oxford: Blackwell.

CHESNEY, R.W., A.J. HAMSTRA, R.B. MAZEES, and H.F. DE LUCA. 1978. Reduction of serum 1,25-dihydroxyvitamin D in children receiving glucocorticoids. *Lancet* 2:1123.

CLEMENTS, J.E. and B.B. ANDERSON. 1980. Glutathione reductase activity and pyridoxine (pyridoxamine) phosphate oxidase activity in the red cell. *Biochem. Biophys. Acta* 632: 159.

COFFEY, G. and C.W.M. WILSON. 1975. Ascorbic acid deficiency and aspirin-induced hematemesis. *Brit. Med. J.* 1: 208.

CRAWFORD, J.D. and G.C. KENNEDY. 1959. Animal physiology: Chlorothiazide in diabetes insipidus. *Nature* 183: 891.

CURTIS, A.C. and R.S. BALLMER. 1939. The prevention of carotene absorption by liquid petrolatum. *J.A.M.A.* 113:1785.

DANIELS, A.L. and G.J. EVERSON. 1936–37. Influence of acetylsalicylic acid (aspirin) on urinary excretion of ascorbic acid. *Proc. Soc. Exp. Biol. Med.* 35: 20.

DE LUCA, H.F. 1980. The control of calcium and phosphorus metabolism by the vitamin D endocrine system. *Ann. N.Y. Acad. Sci.* 355: 1.

DITZLER, J.W. 1970. Epsom salts poisoning and a review of magnesium ion physiology. *Anesthesiology* 32: 378.

DORPH, S. and A. OLGAARD. 1968. Effect of triamterene on serum potassium and serum creatinine in long-term treatment with thiazides. *Nord. Med.* 79: 516.

DRUTZ, D.J., J.H. FAN, T.Y. TAI, J.T. CHENG, and W.C. HSIEH. 1970. Hypokalemic rhabdomyolysis and myoglobinuria following amphotericin B therapy. *J.A.M.A.* 211: 824–26.

DUARTE, C.G. 1968. Effects of chlorothiazide and amipramizide (MK 8780) on the renal excretion of calcium, phosphate and magnesium. *Metabolism* 17: 420.

DUARTE, C.G., J.L. WINNACKER, K.L. BECKER, and A. PAGE. 1971. Thiazide-induced hypercalcemia. *New Engl. J. Med.* 284: 828–30.

DYCKNER, T. and P-O. WESTER. 1985. Renal excretion of electrolytes in patients on long-term diuretic therapy for arterial hypertension and/or congestive heart failure. *Acta Med. Scand.* 218: 443–48.

EICHNER, E.R. and R.S. HILLMAN. 1973. Effect of alcohol on serum folate level. *J. Clin. Invest.* 52: 584.

EKNOYAN, G., W.N. SUKI, and M. MARTINEZ-MALDONADO. 1970. Effect of diuretics on urinary excretion of phosphate, calcium and magnesium in thyroparathyroidectomized dogs. *J. Lab. Clin. Med.* 76: 25.

FLEAR, C.T.G., G.V. GILL, and J. BURNS. 1981. Hyponatremia: mechanisms and management. *Lancet* 2: 26.

FLEMING, B.J., S.M. GENUTH, A.B. GOULD, and M.D. KAMKIONKOWSKI. 1975. Laxative-induced hypokalemia, sodium depletion and hyperreninemia. Effects of potassium and sodium replacement on the renin-angiotensin-aldosterone system. *Ann. Intern. Med.* 83: 60–62.

FLINK, E.B. 1985. Hypomagnesemia in the patient receiving digitalis. *Ann. Intern. Med.* 145: 625–26.

FRANKLIN, J.L. and I.H. ROSENBERG. 1973. Impaired folic acid absorption in inflammatory bowel disease: Effects of salicylazosulfapyridine (Azulfidine). *Gastroenterology* 64: 517.

GABRIEL, R. 1977. *Renal Medicine.* 206. London: Bailliere Tindall.

GAL, L., C. PARKINSON, and I. KRAFT. 1971. Effect of oral contraceptives on human plasma vitamin A levels. *Brit. Med. J.* 2: 436.

GALTEAU, M.M. and O. SIEST. 1984. Drug effects in clinical chemistry. Part 2. Guidelines for evaluation of analytical interference. IFCC Document Stage 2, Draft 3, 1983-11 with a proposal for an IFCC Recommendation. *Clin. Chim. Acta.* 139: 223F–230F.

GERBER, A., A.P. RABB, and A.E. SOBEL. 1954. Vitamin A poisoning in adults. *Am. J. Med.* 16: 729–45.

GERBER, L.H., R.K. HELFGOTT, E.G. GROSS, J.E. HICKS, S.S. ELLENBERG, and G.L. PECI. 1984. Vertebral abnormalities associated with synthetic retinoid use. *J. Am. Acad. Dermatol.* 10: 817–23.

GLUECK, C.J., S. FORD, D. SCHEEL, and P. STEINER. 1972. Colestipol and cholestyramine resin. Comparative effects in familial type II hyperlipoproteinemia. *J.A.M.A.* 222: 676.

GOLDSMITH, R.S. and S.H. INGBAR. 1966. Inorganic phosphate treatment of hypercalcemia of diverse etiologies. *New Engl. J. Med.* 174: 1.

GREVESON, G., M. LYE, C.P. PETCH, and W.H. JOPLING. 1981. Long-term thiazide diuretics. *Lancet* 1: 941.

GRONERT, G.A. 1970. Potassium response to succinylcholine. Letter. *J.A.M.A.* 211: 300.

GROSS, E.G., J.D. DEXTER, and R.G. ROTH. 1966. Hypokalemic myopathy with myo-globinuria associated with licorice ingestion. *New Engl. J. Med.* 274: 602.

GUNGDOGDU, A.S., P.M. BROWN, S. JUUL, L. SACHS, and P.H. SONKSEN. 1979. Comparison of the hormonal and metabolic effects of salbutamol infusion in normal subjects and in insulin-requiring diabetics. *Lancet* 2: 1317–21.

HAGEN, G.A. and T.F. FRAWLEY. 1970. Hyponatremia due to sulfonylurea compounds. *J. Clin. Endocrinol. Metab.* 31: 570.

HAHN, T.J., B.A. HENDIN, C.R. SCHARP, and J.G. HADDAD. 1972. Effect of chronic anticonvulsant therapy on serum 25-hydroxycalciferol levels in adults. *New Engl. J. Med.* 287: 900.

HAMDY, R.C., J. TOVEY, and N. PERERA. 1980. Hypokalemia and diuretics. *Brit. Med. J.* 1: 1187.

HANSTEN, P.D. 1973. *Drug Interactions,* 2nd ed. p. 257. Philadelphia: Lea and Febiger.

HASHIM, S.A., S.S. BERGEN, Jr., and T.B. VAN ITALLIE. 1961. Experimental stea-torrhea induced in man by bile acid sequestrant. *Proc. Soc. Exp. Biol. Med.* 106: 173.

HEIZER, W.D., A.L. WARSHAW, and L. LASTER. 1968. Protein-losing enteropathy and malabsorption associated with factitious diarrhea. *Ann. Intern. Med.* 68: 839.

HERBERT, L.A., J. LEMANN, Jr., J.R. PETERSON, and E.J. LENNON. 1966. Studies on the mechanism by which phosphate infusion lowers serum calcium concentration. *J. Clin. Invest.* 45: 1886.

HERMAN, E. and J.P. RADO. 1966. Fatal hyperkalemic paralysis associated with spi-ronolactone. *Arch. Neurol.* 15: 74.

INSOGNA, K.L. et al. 1980. Osteomalacia and weakness from excessive antacid inges-tion. *J.A.M.A.* 244: 2544.

JELIC-IVANOVIC, Z., S. SPASIC, N. MAJKIC-SINGH, and P. TODOROVIC. 1985. Effects of some anti-inflammatory drugs on 12 blood constituents. Protocol for the study of in vivo effects of drugs. *Clin. Chem.* 31: 1141–43.

- JOPLING, W.H. 1981. Long-term thiazide diuretics. *Lancet* 1: 941.

KOCH-WESER, J. 1976. Acetaminophen. *New Engl. J. Med.* 295: 1297.

KRUMDIECK, C. 1978. A long term study of the excretion of folate and pterins in a human subject after ingestion of ^{14}C folic acid, effect of diphenylhydantoin in admin-istration. *Am. J. Clin. Nutr.* 31: 88.

LAWRENCE, V.A., J.E. LOEWENSTEIN, and E.R. EICHNER. 1984. Aspirin and folate binding: In vivo and in vitro studies of serum binding and urinary excretion of en-dogenous folate. *J. Lab. Clin. Med.* 103: 944–48.

LEKLEM, J.E., D.P. ROSE, and R.R. BROWN. 1973. Effect of oral contraceptives on urinary metabolite excretion after administration of L-tryptophan or L-kynurenine sulfate. *Metabolism* 22: 1499.

LETELLIER, G. and F. DESJARLAIS. 1985a. Analytical interference of drugs in clinical chemistry. 1. Study of 20 drugs on 7 different instruments. *Clin. Biochem.* 18: 345–51.

LETELLIER, G. and F. DESJARLAIS. 1985b. Analytical interference of drugs in clinical chemistry: II—The interference of 3 cephalosporins with the determination of serum creatinine concentration by the Jaffe reaction. *Clin. Biochem.* 18: 352–56.

LEVINE, D., A.W. GOODE, and D.L. WINGATE. 1981. Purgative abuse associated with reversible cachexia, hypogammaglobulinemia and finger clubbing. *Lancet* 1: 919.

LEVINE, R.A. 1968. Steatorrhea induced by para-aminosalicylic acid. *Ann. Intern. Med.* 68: 1265.

LEWIS, J.R. 1974. Carbenoxolone sodium in the treatment of peptic ulcer. A review. *J.A.M.A.* 229: 460.

LIM, P. and E. JACOB. 1972. Magnesium deficiency in patients on long term diuretic therapy for heart failure. *Brit. Med. J.* 2: 620.

LONGENECKER, J.B. and S.G. BASU. 1965. Effects of cholestyramine on absorption of amino acids and vitamin A in man. *Fed. Proc.* 24: 375.

LOTZ, M., E. ZISMAN, and F.C. BARTTER. 1968. Evidence for a phosphorus-depletion syndrome in man. *New Engl. J. Med.* 278: 409.

LUETHI, A. and H. STUDER. 1969. Antidiuretic action of chlorpropamide and tolbutamide. *Minn. Med.* 52: 33.

LUFT, F.C., V. PATEL, M.N. YUM, and S.A. KLEIT. 1976. Nephrotoxicity of cephalothin-gentamycin combinations in rats. *Antimicrob. Agents Chemother.* 9: 831.

LUZECKY, M.H., K.D. BURMAN, and E.R. SCHULTZ. 1974. The syndrome of inappropriate secretion of antidiuretic hormone associated with amitriptyline administration. *South. Med. J.* 67: 495.

MAVROMATIS, F. 1965. Tetracycline nephropathy. Case reports with renal biopsy. *J.A.M.A.* 193: 191.

MAZZE, R.I. and M.J. COUSINS. 1973. Combined nephrotoxicity of gentamycin and methoxyflurane anesthesia. A case report. *Br. J. Anesthesiol.* 45: 394.

McGUIGAN, J.E. 1980. A consideration of the adverse effects of cimetidine. *Gastroenterology* 80: 181.

MESTMAN, J.H., D.S. POCOCK, and A. KIRCHNER. 1969. Lactic acidosis with recovery in diabetes mellitus on phenformin therapy. *Calif. Med.* 111: 181–85.

MILNE, M.D. 1974. Drug interactions and the kidney. In *Clinical Effects of Interaction Between Drugs.* ed. L.E. Cluff and J.C. Petrie. p 193. New York: American Elsevier.

MORENO, M., C. MURPHY, and C. GOLDSMITH. 1969. Increase in serum potassium resulting from the administration of hypertonic mannitol or other solutions. *J. Lab. Clin. Med.* 73: 291–98.

MORGAN, D.B. and C. DAVIDSON. 1980. Hypokalaemia and diuretics: An analysis of publications. *Brit. Med. J.* 1: 905.

MOSES, A.M., J. HOWANITZ, M. VAN GEMERT, and M. MILLER. 1973. Clofibrate-induced antidiuresis. *J. Clin. Invest.* 52: 535.

MOSES, A.M. and M. MILLER. 1974. Drug-induced dilutional hyponatremia. *New Engl. J. Med.* 291: 1234.

NARDONE, D.A., W.J. McDONALD, and D.E. GIRARD. 1978. Mechanisms in hypokalemia. Clinical correlations. *Medicine* 57: 435.

NICHOLLS, M.G., E.A. ESPINER, H. IKRAM, and A.H. MASLOWSKI. 1980. Hyponatremia in congestive heart failure during treatment with captopril. *Brit. Med. J.*
281: 909.

NIELSON, J. 1964. Magnesium-lithium studies. I. Serum and erythrocyte magnesium
in patients with manic states during lithium treatment. *Acta Psychiatr. Scand.* 40:
190.

NUTRITION REVIEWS. 1973. Editorial. Oral contraceptives and vitamin B_6. *Nutr. Rev.*
31: 49.

O'BRIEN, I.A., J. FITZGERALD-FRAZER, I.G. LEWIN, and R.J. CORRALL. 1981.
Hypokalemia due to salbutamol overdosage. *Brit. Med. J.* 282: 1515–16.

OLDAM, R.K. and T.C. POMEROY. 1972. Vincristine-induced syndrome of inappropriate secretion of antidiuretic hormone. *South. Med. J.* 6: 1010.

ORWOLL, E.S. 1982. The milk-alkali syndrome: Current concepts. *Ann. Intern. Med.*
97: 242–48.

PARKE, D.V. and C. IOANNIDES. 1981. The role of nutrition in toxicology. *Ann. Rev.
Nutr.* 1: 207.

PEDERSON, E.B. and H. J. KORNERUP. 1976. Relationship between plasma aldosterone concentration and plasma potassium in patients with essential hypertension
during alprenolol treatment. *Acta Med. Scand.* 200: 263.

PEDERSON, O.L. and E. MIKKELSEN. 1979. Serum potassium and uric acid changes
during treatment with timolol alone and in combination with a diuretic. *Clin. Pharmacol. Ther.* 26: 339.

PITTSLEY, R.A. and F.W. YODER. 1983. Hyperostosis. Skeletal toxicity associated with
long-term administration of 13-cis retinoic acid for refractory ichthyosis. *New Engl.
J. Med.* 308: 1012–14.

PRICE, J.M., M.J. THORNTON, and L.M. MUELLER. 1967. Tryptophan metabolism
in women using steroid hormones for ovulation control. *Am. J. Clin. Nutr.* 20: 452.

RADO, J.P. 1973. Water intoxication during carbamazepine treatment. *Brit. Med. J.* 3:
479.

RAMSEY, L.E., P. BAYLE, and M.H. RAMSAY. 1977. Factors influencing serum potassium in treated hypertension. *Quart. J. Med.* 46: 401.

REYNOLDS, E.H., G. MILNER, D.M. MATTHEWS, and I. CHANARIN. 1966. Anticonvulsant therapy, megaloblastic haemopoiesis, and folic acid metabolism. *Quart. J.
Med.* 35: 521–37.

RIVERS, J.M. and M.M. DEVINE. 1972. Plasma ascorbic acid concentrations and oral
contraceptives. *Am. J. Clin. Nutr.* 26: 684.

ROCK, A.G., F. DECARY, and R.S. COLE. 1981. Orange plasma from tanning capsules.
Lancet 2: 1419.

ROE, D.A. 1974. Effects of drugs on nutrition. Minireview. *Life Sci.* 15: 1219.

_____. 1976. *Drug-Induced Nutritional Deficiencies.* Westport, CT: AVI.

_____. 1977. Nutrition and the contraceptive pill. In *Nutritional Disorders of American
Women.* ed. M. Winick. 37–49. New York: Wiley.

_____. 1979. Interactions between drugs and nutrients. *Med. Clin. N. Amer.* 63: 985.

_____. 1981. Drug interference with the assessment of nutritional status. Symposium on Laboratory Assessment of Nutritional Status. *Clin. Lab. Med.* 1 (4): 647.

_____. 1984. Therapeutic significance of drug-nutrient interactions in the elderly. *Pharm. Rev.* 36: 109S–119S.

ROSE, D.P. 1966. Excretion of xanthurenic acid in the urine of women taking progestin-oestrogen preparations. *Nature* 210: 196.

ROSE, D.P., R. STRONG, J. FOLKARD, and P.W. ADAMS. 1973. Erythrocyte amino-transferase activities in women using oral contraceptives and the effect of vitamin B_6 supplementation. *Am. J. Clin. Nutr.* 26: 48.

ROTH, F. and H. WUTHRICH. 1969. The clinical importance of hyperkalemia following suxamethonium administration. *Brit. J. Anaesthesiol.* 41: 311.

SCHILSKY, R. and T. ANDERSON. 1979. Hypomagnesemia secondary to cisdiammine-chlorplatinum II administration. *Ann. Intern. Med.* 90: 929.

SCHWARZ, K.B., P.D. GOLDSTEIN, J.L. WITZTUM, and G. SCHONFELD. 1980. Fat soluble vitamin concentrations in hypercholesterolemic children treated with coles-tipol. *Pediatrics* 65: 243–50.

SHERWOOD, L.M. 1967. Hypernatremia during sodium sulfate therapy. Letter. *New Engl. J. Med.* 277: 314.

SHOJANIA, A.M., G. HORNADY, and P.H. BAINES. 1971. The effect of oral contra-ceptives on folate metabolism. *Am. J. Obstet. Gynecol.* 111: 782.

SMITH, M.J.H. and P.K. SMITH, Eds. 1966. *The Salicylates: A Critical Bibliographic Review.* New York: Interscience.

SPENCER, H., J. GREENBERG, E. BERGER, M. PERRONE, and D. LASZLO. 1956. Studies on the effect of ethylenediaminetetraacetic acid in hypercalcemia. *J. Lab. Clin. Med.* 47: 29–41.

SPENCER, H. and I. LEWIN. 1963. Derangements of calcium metabolism in patients with neoplastic bone involvement. *J. Chronic Dis.* 16: 713.

SPOONER, R.J., R.A. PERCY, and A.G. RUMLEY. 1979. The effect of erythrocyte ageing on some vitamin and mineral dependent enzymes. *Clin. Biochem.* 12: 289.

SRIVASTAVA, L.S., E.E. WERK, K. THRASHER, L.J. SHOLITON, R. KOZERA, W. NOLTEN, and H.C. KNOWLES. 1973. Plasma cortisone concentration as measured by radioimmunoassay. *J. Clin. Endocrinol. Metab.* 36: 937.

STARK, J.J. and S.B. HOWELL. 1978. Nephrotoxicity of cisplatinum (II) dichlorodiam-mine. *Clin. Pharmacol. Thera.* 23: 461.

STEBBINS, R., J. SCOTT, and V. HERBERT. 1973. Drug-induced megaloblastic ane-mias. *Sem. Hematol.* 10: 235.

STEELE, T.H., A.A. SEIPIEK, and J.B. BLOCK. 1973. Antidiuretic response to cyclo-phosphamide in man. *J. Pharm. Exp. Therap.* 185: 245.

SUKI, W.N., J.J. YIUM, M. VON MINDEN, C. SALLER-HERBERT, G. EKNOYAN, and M. MARTINEZ-MALDONADO. 1970. Acute treatment of hypercalcemia with furosemide. *New Eng. J. Med.* 283: 836.

SULLIVAN, J.F., H.G. LANKFORD, M.J. SWARTZ, and C. FARRELL. 1963. Magnesium metabolism in alcoholism. *Am. J. Clin. Nutr.* 13: 297–303.

THORN, G.W. 1966. Clinical considerations in the use of corticosteroids. *New Engl. J. Med.* 274: 775.

THURNHAM, D.I. 1981. Red cell enzyme tests of vitamin status: Do marginal deficiencies have any physiological significance? *Proc. Nutr. Soc.* 40: 155.

WACKER, W.E.C. and A.F. PARISI. 1968. Magnesium metabolism. *New Engl. J. Med.* 278: 712, 772.

WALDENLEND, L., S. BERG, and B. VIKANDER. 1980. Effect of per oral thiamine treatment on thiamine contents and transketolase activity of red blood cells in alcoholic patients. *Acta Med. Scand.* 209: 209.

WAXMAN, S. 1979. The value of measurement of folate levels by radioassay. In *Folic Acid in Neurology, Psychology and Internal Medicine.* ed M.I. Boter and E.H. Reynolds. 47. New York: Raven Press.

WEST, R.J. and J.K. LLOYD. 1975. The effect of cholestyramine on intestinal absorption. *Gut* 16: 93.

WHANG, R., T.O. OEI, and A. WATANABE. 1985. Frequency of hypomagnesemia in hospitalized patients receiving digitalis. *Arch. Intern. Med.* 145: 655–56.

WICKRAMASINGHE, S.N. and J.E. SAUNDERS. 1977. Results of three years' experience with the deoxyuridine suppression test. *Acta Haematol.* 58: 193.

WILBERG, J.J., G.G. TURNER, and F.Q. NUTTALL. 1978. Effect of phosphate or magnesium cathartics on serum calcium. Observations in normocalcemic patients. *Arch. Intern. Med.* 138: 1114.

YUI, Y., Y. ZLOKOWA, and C. KAWAI. 1980. Furosemide-induced thiamine deficiency. *Cardiovasc. Res.* 14: 537.

Chapter 9

PREDICTION AND PREVENTION OF DRUG–NUTRIENT INTERACTIONS USING THEORETICAL MODELS

LIMITATIONS OF CONVENTIONAL RISK ASSESSMENT

Until the present time, drug–nutrient interactions have been predicted on the basis of past experience. Indeed, the risk has been considered from the standpoint of the drugs used rather than user characteristics and conditions of usage. The need for better methods of risk assessment is related to our need to predict when drug–nutrient interactions are likely to occur and how to prevent adverse outcomes. We would like to be able to forecast interactions of new drugs from our knowledge of drugs in current usage. Currently we have a knowledge base which includes the attributes of drugs, drug users, and dosage, but we have not established the interrelationships between these variables.

SIMPLE MATHEMATICAL MODELS

A number of authors have proposed the use of simple mathematical models to elucidate clinical problems (Feinstein 1967; Ledley and Lusted 1959; Wulff 1976a, 1976b). For example, Wulff (1976b) provided solutions to clinical conundrums using elementary set theory. Venn diagrams are used to characterize groups of patients with and without specific symptoms or signs of disease. This same system can be used to analyze drug–nutrient interactions, particularly when the aim is to define the population at risk. For example, while it is now well recognized that a number of drugs cause aberrations in the normal metabolism of vitamin D and induce secondary calcium malabsorption, the task of the physician is to be able to predict which patients taking these drugs are actually at risk for development of rickets or osteomalacia.

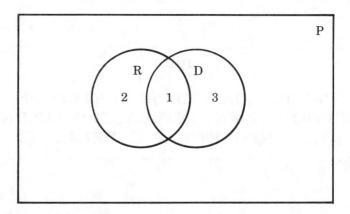

Figure 9.1. Venn diagram showing attributes of drug-taking and non-drug patient groups with and without rickets. P = patient population; R = patients with rickets; D = patients taking a calcium-depleting drug. Subset 1 = R ∩ D = patients taking a calcium-depleting drug (1) who have rickets; 2 = R ∩ D̄ = patients with rickets not taking the drug; 3 = R̄ ∩ D = patients taking the drug who do not have rickets.

The first task, which is to conceptualize this problem, can be achieved by use of Venn diagrams, according to Wulff's (1976b) recommendations. Thus, we can define within our total patient population that subgroup—taking drugs which interfere with vitamin D metabolism—which will develop rickets or osteomalacia by knowing the subsets of the total population. There is a subset with rickets or osteomalacia not taking such drugs, but having environmental or disease-related and/or diet-related factors that explain their bone disease. There is a further subset of patients taking the drugs in question, who do not have rickets or osteomalacia. We can see, then, that there is an overlap between these subsets and that the patients taking the drugs that cause calcium depletion and having rickets or osteomalacia are, in fact, those who take the drugs and have other etiological factors that place them at risk for rickets or osteomalacia. These relationships are explained in the Venn diagrams shown in Figures 9.1 and 9.2.

This process can be refined further to determine the specificity and sensitivity of knowledge of drug usage as a determinant of rickets (or osteomalacia). How to gather information for our knowledge base (concerning the drug and drug user), and the prevalence of disease on the predictive value of this information is shown in Figures 9.3 and 9.4 with the examples. The best prognostic strategy in relation to drug-induced malnutrition may be to select nosographic characteristics of the drugs in question (e.g., calcium-depleting potential) and users with

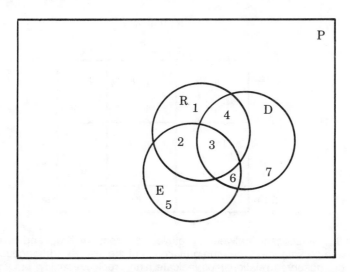

Figure 9.2. Venn diagram showing interacting drug and environmental determinants of rickets. Subset 1 = patients with rickets but without drug-related or environmentally related "cause" $R \cap \bar{D} \cap \bar{E}$ (lack of sunlight); 2 = patients with rickets and environmentally related cause but not taking the calcium-depleting drug $R \cap E \cap \bar{D}$; 3 = patients with rickets and environmentally related "cause" who are taking the drug $R \cap D \cap E$; 4 = patients with rickets who are taking the drug who do not have environmentally related "cause" $R \cap D \cap \bar{E}$; 5 = patients with environmentally related "cause" for rickets but who do not actually have the disease and are not taking the drug $\bar{R} \cap \bar{D} \cap E$; 6 = patients with environmentally related "cause" and are taking the drug who do not have rickets $\bar{R} \cap D \cap E$; 7 = patients taking the drug who do not have rickets and who do not have environmentally related cause $\bar{R} \cap D \cap \bar{E}$.

the highest specificity and sensitivity for outcome (e.g., those at risk for rickets due to indoor environment) (see Figures 9.3 and 9.4). Unfortunately, because of the multifactorial nature of most adverse outcomes of drug–nutrient interactions, these simplistic models seldom meet our need to define risk groups.

It is indeed essential to consider combinations of nosographic characteristics of the drug, the drug regimen, and the drug user that best define the prognosis.

KNOWLEDGE BASE MANAGEMENT (EXPERT SYSTEMS)

Instead of playing the game of trial and error or using a conventional probability approach, we can now use a knowledge base management or "expert" system, which will enable us to make a logical selection of

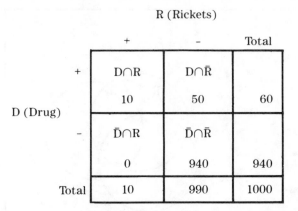

Figure 9.3. Relationship between the etiological agent (the drug) and the disease (rickets) with prevalence. When the prevalence of the disease (drug-induced rickets) is high, observed relationships indicate high predictive ability, but when the prevalence is low or very low, which is usual in a U.S. population, the predictive value of knowing about the drug intake is low for the disease (see Example 1 in Table 9.1).

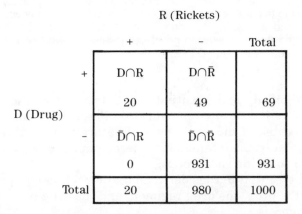

Figure 9.4. Relationship between the etiological agent (the drug and the environment of the patient) and the disease with prevalence. When the prevalence of the disease is low in the total patient population (H) but high in those who have other determining etiological factors for the disease (R), then we can examine this subpopulation to see the predictive ability of knowing about drug usage (usage of calcium-depleting drugs [D]), and we see that the predictive ability is high (see Example 2).

nosographic characteristics of prognostic importance. Such knowledge base management systems have previously been used to search, sort, summarize, or process biomedical data for medical research and teaching. When such systems have been used for diagnostic or prognostic purposes, they have had the following components:

- A collection of facts (data base)
- A set of production rules
- A knowledge and inference structure
- A mechanism for drawing inferences from incomplete evidence

Some of these systems have been developed by the collaborative efforts of physicians and computer scientists (Weiss et al. 1978). However, due to the complexity of producing such general problem-solving methods, recent emphasis has been on automating problem-solving in specific areas using defined or domain-specific knowledge bases (Alty and Coombs 1984). Expert systems that may be required to investigate drug–nutrient interactions include:

- Differential diagnosis systems
- Risk-assessment systems
- Rule-generating systems
- Rule-testing systems
- Information-gathering systems
- Process guides

Differential diagnosis systems serve to discriminate or rule out alternate explanations of the signs of chemically induced disease. Risk-assessment systems use comparisons of attributes or logistic risk determinations. Rule-generating systems may be drug-related or related to the drug user or regimen. Rule-testing systems may be related to "new" drugs or drug users. Information-gathering systems expand existing data bases. Process guides are used to explain appropriate intervention in given circumstances.

A glossary of terms is needed by the new user of expert systems. Terms in current usage are as follows:

- Domain: a specific knowledge base
- Description of the domain: documentation of the knowledge base
- Elements: items or groups within the domain, e.g., drugs
- Attributes: characteristics of the elements
- Attribute values: qualitative or quantitative scales that classify the attributes, e.g., nominal, ordinal, interval, or logical values.

We can use an expert system for the following purposes related to drug–nutrient interactions:

TABLE 9.1 EXAMPLE 1: PREDICTIVE VALUES OF A POSITIVE
INDICATOR WITH 99% SENSITIVITY AND 95% SPECIFICITY
AT TWO LEVELS OF PREVALENCE

	1%	2%
Number of population	1000	1000
Number with rickets	10	20
Number without rickets	990	980
Number with rickets using drug	10	20
Number without rickets using drug	50	49
Total number using the drug	60	69
Predictive value of information on drug usage	$\dfrac{10}{60} = 17\%$	$\dfrac{20}{69} = 29\%$

- Predicting incompatibility reactions
- Predicting change in drug or nutrient bioavailability
- Predicting change in the rate of drug metabolism in response to dietary change
- Defining drug user groups at risk for drug-induced malnutrition
- Determining "new" drug outcomes related to food, nutrient, or alcohol interactions
- Providing rational process guides for health professionals

PRACTICAL USE OF AN EXPERT SYSTEM

In the following examples we will use an expert system in the exploration of the probability of adverse outcomes from "new" and "old" drugs. "Expert 4" is a software program for microcomputers[1] that can be used to predict when drug–nutrient interactions are likely to occur. With this system, *domains* of reactions as well as *elements* (which may be drugs, foods, formulas, or supplements) are defined. *Attributes* are drug, food, formula, and nutrient supplement properties that are specific to drugs or foods, and these attributes can be classified by *attribute values*. Matrices can be generated that relate the attributes and attribute values of the elements. The similarities and differences of the attributes and attribute values can then be compared. Typical characteristics of drugs having different values on particular attributes

[1]Expert 4 was supplied by the author under License No. 0018 by the biomedical distributor Elsevier Biosoft and was developed by R. Rivers, G.5 Expert Systems, 18 Sarah House, Arabella Drive, London, SW15, 5LN, England.

TABLE 9.2 ELEMENTS IN DOMAIN "DRUG COMPATIBILITY WITH ENTERAL FORMULAS" ("DRUG pH")

Donnatal elixir (Robbins)
Docusate sodium (DDS)
Lomotil liquid (Searle)
Mylicon drops (Stuart)
Phenobarb elixir (Parke-Davis)
Actifed syrup (Burroughs-Wellcome)
Dimetane elixir (Robbins)
Dimetapp elixir (Robbins)
Phenergan syrup (Wyeth)
Sudafed syrup (Burroughs-Wellcome)
Cibalith-S syrup (Ciba)
Mellaril oral (Sandoz)
Bactrim ds susp. (Roche)
Feosol elixir (Menley and James)
KCl liquid (Barre)
Klorvess syrup (Dorsey)

will be identified. Attribute cross-tabulations (e.g., drug properties or user groups) will be set out in frequency distributions so that we can generate hypotheses concerning the relationship between any two attributes. Searches can be made for elements which are similar in attributes. Attribute statistics can be generated. The probability that knowledge about the attribute values of an "old" (commonly used) drug will determine effects of a "new" drug, on the basis of its attribute values, can be evaluated. An index of the probability for immediate classification (PIC index) can be used with a full domain to find out the good predictor attributes within that model and to generate hypotheses about the relationships between attribute values. A search profile can be used as a criterion for inclusion or exclusion of particular attributes that are of prognostic significance in specific drug–nutrient interactions.

Example: Drug Compatibility with Enteral Formulas. In this domain, the elements are drugs (liquid drug formulations) and the formulas are Ensure Plus, Ensure, and Osmolite. The example is based on the biopharmaceutical study reported by Cutie et al. (1983). The elements (16 products) are listed in Table 9.2. The elements have three attributes: pH of the product, pH of the mixture, and compatibility of the drug and the formulas. Each attribute has attribute values. The attribute values for pH of the product and of the mixture are in pH ranges

TABLE 9.3 DISTRIBUTION OF VALUES ON THE ATTRIBUTE:
pH PRODUCT

Value	Quantity	A Priori	Profile
Unknown	2	0.111	****
<2.5	2	0.111	****
2.5–2.9	3	0.167	*******
3.0–3.4	1	0.056	**
3.5–3.9	2	0.111	****
4.0–4.4	0	0	
4.5–4.9	3	0.167	*******
5.0–5.4	1	0.056	**
5.5–5.9	2	0.111	****
6.0–6.4	1	0.056	**
6.5–6.9	1	0.056	**

(see Tables 9.3 to 9.5). The attribute values of compatibility are compatibility (C) and incompatibility (I). The attribute values are tabulated in relation to 16 drug formulations in Table 9.6. Differences between products can then be examined (e.g., differences between Dimetane Elixir and Phenergan Syrup include the pH of the product, the pH of the mixture, and the compatibility of the mixture [see Table 9.7]). The probability of each attribute classifying the class attribute (PIC index), which in the selected case is compatibility, can be tested (see Table 9.8). The PIC index shows the percentage of elements that each attribute will immediately classify within the class attribute. It is constructed out of individual probabilities that are added across attribute values for each attribute. It is, however, valid only for the

TABLE 9.4 DISTRIBUTION OF VALUES ON THE ATTRIBUTE:
pH MIXTURE

Value	Quantity	A Priori	Profile
Unknown	2	0.111	****
<3.0	0	0	
3.0–3.4	2	0.111	****
3.5–3.9	0	0	
4.0–4.4	1	0.056	**
4.5–4.9	2	0.111	****
5.0–5.4	2	0.111	****
5.5–5.9	3	0.167	*******
6.0–6.4	3	0.167	*******
6.5–6.9	3	0.167	*******

TABLE 9.5 DISTRIBUTION VALUES ON THE ATTRIBUTE: COMPATIBILITY

Value	Quantity	A Priori	Profile
Unknown	2	0.111	****
C	8	0.444	*****************
I	8	0.444	*****************

TABLE 9.6 MAIN ATTRIBUTE VALUE TABLE (16 ELEMENTS, 3 ATTRIBUTES)

	pH Products	pH Mixtures	Compatibility
Donnatal elixir	4.5–4.9	5.0–5.4	C
Docusate sodium	6.0–6.4	6.5–6.9	C
Lomotil liquid	3.0–3.4	6.5–6.9	C
Mylicon drops	4.5–4.9	6.0–6.4	C
Phenobarb elixir	6.5–6.9	6.5–6.9	C
Actifed syrup	5.5–5.9	6.0–6.4	C
Dimetane elixir	2.5–2.9	4.4–4.9	I
Dimetapp elixir	2.5–2.9	5.0–5.4	I
Phenergan syrup	5.0–5.4	5.5–5.9	C
Sudafed syrup	2.5–2.9	4.5–4.9	I
Cibalith-S syrup	4.5–4.9	4.5–4.9	I
Mellaril oral	3.5–3.9	5.5–5.9	I
Bactrim ds susp.	5.5–5.9	6.0–6.4	C
Feosol elixir	<2.5	3.0–3.4	I
KCl liquid	3.5–3.9	5.5–5.9	I
Klorvess syrup	<2.5	3.0–3.4	I

TABLE 9.7 DIFFERENCES BETWEEN DIMETANE ELIXIR (ROBBINS) AND PHENERGAN SYRUP (WYETH)

	Dimetane	Phenergan
pH product	2.5–2.9	5.0–5.4
pH mixture	4.0–4.4	5.5–5.9
Compatibility	I	C

TABLE 9.8 PROBABILITY OF EACH ATTRIBUTE CLASSIFYING THE
CLASS ATTRIBUTE PROBABILITY

pH product	74
pH mixture	62

model in question. In the present example, the PIC index tells us that knowledge of the pH of the product is most important in predicting whether or not the drug and formula will be compatible. Conversely, when it is reported that a nasogastric tube became blocked when a liquid preparation of a drug has been administered during an enteral formula infusion, then it is likely that the drug product is, in fact, acidic.

Example: Prediction and Avoidance of Food and Over-the-Counter (OTC) Drug Interference with Tetracycline Bioavailability. In this example, the first aim is to predict when the absorption and resultant therapeutic effect of tetracycline will be reduced by ingestion of a food or OTC drug product. Secondly, the need is to find out when meals should be eaten and OTC drugs taken to avoid loss of the desired therapeutic effect of the tetracycline. The elements are food (including dairy products, cereal, and vegetables), and OTC drug groups (including antacids and analgesics). The attributes are food and OTC drug classification and effects of ingestion on plasma drug (tetracycline) level as well as when the effects occur ("When"). The attribute values of foods and drugs are "cation-rich" or "cation-poor." (Cations known to reduce the bioavailability of tetracycline are calcium, magnesium, aluminum, iron, and zinc, which form chelates with the antibiotic). The attribute values of *effects* on drug level include those related to presence or absence of effect. They are designated "reduced," meaning reduced blood level of the drug, or "adequate," meaning adequate blood level. Those related to when food–drug or drug–drug interactions occur are ordinal and are designated "<2h" and ">2h," meaning that less than or greater than two hours' interval between food or OTC drug and tetracycline is associated with an effect. In Tables 9.9 and 9.10, we can see that reduced levels of tetracycline can be predicted from knowledge of ingestion of "cation-rich" foods or OTC drugs, and when less than two hours elapses between intake of these items and the tetracycline. We therefore know that if the drug is to be taken every six hours, the "cation-rich" items must be taken more than two hours away from the tetracycline times to prevent lowered bioavailability. The optimal timing of food/OTC drug intakes in relation to tetracycline is shown in Table 9.11.

TABLE 9.9 RELATIONSHIP BETWEEN INTAKE OF "CATION-RICH" FOOD OR DRUGS AND REDUCED BLOOD LEVEL OF TETRACYCLINE

	Class	Effect	When
Dairy	Cation +	Reduced	<2h
Cereal	Cation −	Adequate	<2h
Vegetables	Cation −	Adequate	<2h
Antacids	Cation +	Reduced	<2h
Analgesics	Cation −	Adequate	<2h

Example: Warfarin Resistance and Enteral Formulas. Ingestion of vitamin K–rich foods or formulas can,lead to warfarin resistance in patients receiving such coumarin anticoagulants for the treatment of venous thromboses or pulmonary emboli. The aim is to predict when warfarin resistance will occur. The example is based on the report by Howard and Hannaman (1985). The elements are nutrition products (enteral formulas), including Isocal, Ensure, Osmolite, Sustacal, Vivonex, and Meritene. The attributes are vitamin K intake from these sources and change in prothrombin time. The attribute values for vitamin K intake per day from the formulas are ordinal and expressed in micrograms. The attribute values for change in prothrombin time

TABLE 9.10 PROBABILITY OF EACH ATTRIBUTE CLASSIFYING THE CLASS ATTRIBUTE

Effect	
1	Classification of food/OTC drug (cation + vs) = 100
3	When taken = 0

TABLE 9.11 INTERVAL IN HOURS (H) REQUIRED TO AVOID LOW BLOOD TETRACYCLINE LEVELS WHEN DAIRY FOODS OR ANTACIDS ARE TAKEN

	Class	Effect	Avoid
Dairy	Cation +	Reduced	<2h
Cereal	Cation −	Adequate	<2h
Vegetables	Cation −	Adequate	<2h
Antacids	Cation +	Reduced	<2h
Analgesics	Cation −	Adequate	<2h

TABLE 9.12 RELATIONSHIP BETWEEN THE FORMULA, ITS
VITAMIN K CONTENT, AND THE PROTHROMBIN TIME
(SIX ELEMENTS, FIVE ATTRIBUTES)

	Vit. K mcg	Pro. time
Isocal	>100 <500	Decreased
Ensure	>50 <100	Prolonged
Osmolite	>100 <500	Decreased
Sustacal	>100 <500	Decreased
Vivonex	<25	Prolonged
Meritene	<25	Prolonged

are "prolonged" and "decreased." In this example, we can see that it
is possible to predict warfarin resistance on the basis of the formula
ingested and the vitamin K obtained from this source. It is noted that
in the United States, the vitamin K intake from the diet is usually
between 300 and 500 micrograms per day. Relationships between the
elements, their attributes, and the attribute values are shown in Table
9.12. The probability of each attribute classifying the class attribute
"prothrombin time" is shown in Table 9.13.

Example: Risk of Folate Depletion From a New Drug.   In this domain,
the elements are the drugs (phenytoin, theophylline, cholestyramine,
colestipol, niacin, aspirin, and new drugs) and the attributes are mechanism, indication, side effect, patient group, and duration of drug usage.
The attribute values for the five attributes are shown in Tables 9.14
to 9.18. In Table 9.19, we can see the spreadsheet showing the relationship between the drugs, their attributes, and the attribute values.
Knowing that the new drug is a bile-acid sequestrant, which would be
prescribed for prolonged usage in children with familial Type II hyperlipoproteinemia, and knowing that other drugs having the same
mechanism, indication, and risk can induce folate deficiency, we predict
this outcome with the new drug (see Table 9.20), and show importance
of the attribute and attribute values in relation to outcome (see Table
9.21).

TABLE 9.13 PROBABILITY OF EACH ATTRIBUTE CLASSIFYING THE
CLASS ATTRIBUTE PROTHROMBIN TIME

1	Vit. K mcg	100
5		0
4		0
3		0

TABLE 9.14 ATTRIBUTE VALUES FOR MECHANISM

0 Unknown
1 Bile acid sequestrant
2 Analgesic
3 Anticonvulsant
4 Hypolipemic
5 Bronchodilator

Example: Prediction of a Drug-Related Hypertensive Crisis (When to Expect a Tyramine Reaction). Here we will consider the salient points that need to be considered when taking a history from a patient ("John") who has developed a hypertensive crisis after going to a wine-and-cheese party.

Let us use the recall of foods and beverages consumed by all the different people who attended the party and find out if any of them had a similar attack. Several of the other guests drank and ate the same wine and snacks as the patient. We may therefore ask: what is different about the patient? Is a history of previous hypertension rel-

TABLE 9.15 ATTRIBUTE VALUES FOR INDICATION

0 Unknown
1 Epilepsy
2 Asthma
3 Type II hypercholesterolemia
4 Headache

TABLE 9.16 ATTRIBUTE VALUES FOR SIDE EFFECT

0 unknown
1 folate deficiency
2 GI bleeding
3 nausea
4 flush

TABLE 9.17 ATTRIBUTE VALUES FOR GROUP

0 unknown
1 children
2 adults

TABLE 9.18 ATTRIBUTE VALUES FOR DURATION

0 unknown
1 short
2 long

evant? What about drug use? In order to use this case to teach students about acute incompatibility reactions, we will make the party guests the elements. Their attributes will include foods eaten at the party, previous hypertension, drug usage, and whether or not these drugs impose a risk of causing acute hypertension. The attribute values of "foods eaten" will be the party foods and wine, i.e., cheddar cheese, pretzels, and peanuts. The attributes of previous hypertension will be yes and no. The attributes of drug usage will be the drugs taken by the guests. The attribute values of drug-related hypertension risk will also be yes or no. When the main attribute value table is examined (see Table 9.22), it can be seen that guests other than the patient, John, ate the cheese and drank the wine. The other guests, Mary and Peter, had a history of prior hypertension, and Peter took the same drug, an antidepressant monamine oxidase inhibitor, but only John had an attack. The attribute values (etiological factors) that indicate risk of hypertensive crisis (see Table 9.23) can then be defined. If the example is used to teach a class of students, the students can then be asked to explain these findings and to explain why the patient had a tyramine reaction and on what basis they came to this conclusion.

TABLE 9.19 MEAN ATTRIBUTE VALUE TABLE (SEVEN ELEMENTS, FIVE ATTRIBUTES)

	Mechanism	Indication	Side Effect	Group	Duration
Phenytoin	anticonv	Epilepsy	folate def.	children	long
Theophylline	bronchod	Asthma	nausea	children	short
Cholestyramine	bile acid	Type II	folate def.	children	long
Colestipol	bile acid	Type II	folate def.	children	long
Niacin	hypolipe	Type II	flush	adults	long
Aspirin	analgesic	Headache	GI bleeding	children	short
New	bile acid	Type II	unknown	children	long

TABLE 9.20
PROBABILITY OF EACH ATTRIBUTE CLASSIFYING THE CLASS
ATTRIBUTE SIDE EFFECT

1	mechanism	85
2	indication	42
4	group	14
5	duration	0

TABLE 9.21 PROTOTYPE FOR SIDE EFFECT: FOLATE DEFICIENCY

		Association Index	Distinguishability Index
Mechanism	Bile acid sequestrant	Strong	Good
Indication	Type II hyperlipoproteinemia	Strong	Medium
Group	Children	Very strong	Medium
Duration	Long	Very strong	Medium

TABLE 9.22 ATTRIBUTE VALUE COMBINATIONS RELATED TO OUTCOME (RISK OF HYPERTENSIVE CRISIS): MAIN ATTRIBUTE TABLE (SEVEN ELEMENTS, FOUR ATTRIBUTES)

	Foods	Past High B.P.	Drug Usage	Risk of Hypertensive Crisis
Mary	cheese	Yes	Thiazide	No
John	cheese	Yes	MAOI	Yes
Ann	cheese	No	None	No
Peter	pretzels	Yes	MAOI	No
Robert	peanuts	No	None	No
Jane	cheese	No	None	No
Adam	cheese	No	None	No

TABLE 9.23 PROTOTYPE FOR RISK OF ACUTE HYPERTENSION: YES

		Association Index	Distinguishability Index
Foods	cheese	Very strong	Very poor
Past high BP	yes	Very strong	Poor
Drug usage	MAOI	Very strong	Medium

REFERENCES

ALTY, J.L. and M.J. COOMBS. 1984. *Expert Systems: Concepts and Examples*. Manchester, England: NCC.

CUTIE, A.J., E. ALTMAN, and L. LENKEL. 1983. Compatibility of enteral products with commonly employed drug additives. *J. Parent. Enter. Nutr.* 7:186.

FEINSTEIN, A.R. 1967. *Clinical Judgment*. Baltimore: Williams and Wilkins.

HOWARD, P.A. and K.N. HANNAMAN. 1985. Warfarin resistance linked to enteral nutrition products. *J. Am. Dietet. Assoc.* 85: 713–15.

LEDLEY, R.S. and L.B. LUSTED. 1959. Reasoning foundations of medical diagnosis. *Science* 130:9–21.

WEISS, S.M., C.A. KULIKOWSKI, S. AMARAL, and A. SAFIR. 1978. A model-based method for computer-aided medical decision-making. In *Artificial Intelligence*. 148, 166. North Holland Publ. Co.

WULFF, H.R. 1976. *Rational Diagnosis and Treatment: An Introduction to Medical Decision Making*. 1st ed. London and Boston: Blackwell.

———. 1981. *Rational Diagnosis and Treatment: An Introduction to Medical Decision Making*. 2nd ed. London and Boston: Blackwell.

Chapter 10

PROCESS GUIDES FOR DRUG–NUTRIENT INTERACTIONS

Guidelines for drug–nutrient risk assessment and intervention are given in the following process guides. General guidelines for health professionals include:

1 Take a drug history from the patient or, when necessary, from the caregiver.
2 Examine all the prescription (Rx) and over-the-counter (OTC) drugs that the patient is taking and record these on a drug form indicating dose, frequency, duration, and when the drugs are taken in relation to food.
3 In prescribing, avoid multiple drugs when possible.
4 When multiple drugs are required, check to avoid combinations that have adverse nutritional effects.
5 Avoid unnecessary changes during drug therapy, including changes in timing.
6 Take a diet history, including information on alcohol intake and intake of nutrient supplements.
7 Minimize changes in the diet and consult others on the health care team.
8 Observe and record response to drug therapy.
9 Monitor the patient's nutritional status, especially when drug use alters appetite, impairs nutrient absorption or utilization, or promotes nutrient excretion.
10 Inform the patient about possible risks of identifiable drug–food, drug–nutrient and drug–alcohol incompatibilities.

Drug–nutrient interactions are grouped by age group at special risk in Table 10.1. However, in general, the risks of adverse outcomes from drug interactions are similar in other age groups. Enhanced risks are associated with infants with immaturity, children with growth abnor-

177

TABLE 10.1 PROCESS GUIDES RELATED TO COMMON DRUGS USED IN SPECIFIC AGE GROUPS

Infants and Children

Drug	Interaction	Risk	Intervention
Theophylline	Protein	Rapid rate of drug metabolism	Avoid change to high-protein diet that can impair asthma control.
	Milk or infant formula	Slowed rate of drug absorption	Give liquid form of drug in fasting state.
Phenytoin	Vitamin D and calcium	Rickets	Increased sunlight exposure or give UV treatment. Give vitamin D_3 2000–4000 IU/day for 3 months to cure rickets.
	Folate	Megaloblastic anemia	Check CBC and plasma/RBC folate.
		Loss of seizure control	Give folic acid 0.4–1.0 mg/day. Do not give folic acid supplement with drug or >5 mg/day.
	Vitamin K	Hemorrhagic disease of newborn	Give 5 mg vitamin K to mother for 5 days before delivery.
	Enteral formula	Loss of seizure control	Stop flow. Flush nasogastric tube. Give liquid drug preparation. Restart flow 1 hr. post-drug.
Penicillin V	Food	Reduced drug absorption	Give drug fasting or 2 hrs. after food.
Boric acid	Riboflavin	Borate toxicity	Avoid use. If drug toxicity is present, give IV fluids and 10 mg riboflavin/day.

Drug	Interaction	Risk	Intervention
Tetracycline	Calcium	Inadequate drug absorption Impaired calcium absorption	Give drug 2 hr. before or after food/beverage containing calcium.
	Iron	Impaired drug absorption	Do not give drug with iron-rich food or iron supplement.
Isotretinoin	High-fat diet	Elevated plasma triglycerides	Avoid intake of fat calories above 30% of total.
Metronidazole	Alcohol	Disulfiram-like reaction	Avoid alcoholic beverages.
Oral contraceptives (OC)	Folate	Folate depletion	Advise a folate-rich diet. If pregnant within 6 mo. of OC, give folic acid supplement, 0.4–1 mg/day.
	Vitamin B_6	Vitamin B_6 depletion	Give vitamin B_6 supplement, 5 mg/day

Middle-Aged Adults

Drug	Interaction	Risk	Intervention
Chlorpropamide Cimetidine	Alcohol Vitamin B_{12}	Flush reaction Vitamin B_{12} depletion and megaloblastic anemia with long-term use	Avoid alcoholic beverages. Check plasma vitamin B_{12}, RBC, and MCV. Avoid intake with animal protein foods.

Elderly

Drug	Interaction	Risk	Intervention
Warfarin	Vitamin K	Loss of anticoagulant drug effects	Avoid use of formula foods containing vitamin K.

(continued)

TABLE 10.1 PROCESS GUIDES RELATED TO COMMON DRUGS USED IN SPECIFIC AGE GROUPS (CONT.)

Elderly

Drug	Interaction	Risk	Intervention
	Vitamin E	Potentiation of drug effects; bleeding	Stop vitamin E supplementation.
	Cholestyramine, Colestipol	Low warfarin absorption	Stop cholestyramine/colestipol.
Aspirin	Alcohol	GI blood loss and anemia	Stop alcohol and aspirin.
Antacid (Aluminum hydroxide) (Sodium bicarbonate)	Phosphate	Phosphate depletion and osteomalacia	Avoid antacid at mealtimes.
	Folate	Folate deficiency and anemia	Avoid antacid with food folate source.
Thiazide	Digoxin	Digitalis toxicity	Give potassium supplement.
	Lithium	Sodium depletion and lithium toxicity	Stop thiazide.
	Antidiabetic agent	Elevated blood sugar	Stop thiazide.
	Thiazide	Potassium deficiency	Stop laxative and give potassium supplement.
Laxative	Laxative	Laxative abuse can induce malabsorption and osteomalacia.	Stop laxatives. Check stool fat and plasma calcium.
	Mineral oil	Mineral oil with or without laxative can reduce absorption of fat-soluble vitamins.	Stop mineral oil and laxative. Give vitamin A, D, and K supplements.
Antidepressant (MAOI)	Tyramine	MAOI drugs with tyramine in food (cheese) causes high blood pressure.	Give labetalol.

180

mality, young adults using drugs for acne or for birth control, and in the elderly with multiple-drug regimens, special diets, chronic disease, and marginal nutrient intake.

Process guides for drug–nutrient interaction, developed for use in geriatric care institutions, are given later in the chapter.

Drug–nutrient interactions are avoidable. Their prevention and their adverse outcomes are the joint responsibility of the members of the health care team, as follows:

1   *The physician.* Recognition of risk, prediction of adverse outcomes of specific drug and drug–food combinations, nutritional assessment, and monitoring of drug users. Physicians must be aware that a drug's ineffectiveness or toxicity may be related to a food, formula, nutrient supplement, or to unsuitable diet.
2   *The nurse.* Knowledge of when to give medication in relation to food and how to give drugs to patients on enteral feeding regimes.
3   *The dietitian.* Familiarity with drug–nutrient interactions and of dietary changes that can lead to loss of drug efficacy and drug–food incompatibilities. The dietitian is given the specific responsibility to provide process guides for other members of the health care team.
4   *The pharmacist.* Instruction of the patient, appropriate reminding of the physician, and labeling of drugs so the user knows when to take medication in relation to food. Advice on nutrient supplements that may interfere with the desired therapeutic effect of drugs.

Predicting the risk of drug–nutrient interactions and prevention is, in principle, a science, not an art. *A priori* prediction and prevention in every case, however, requires knowledge of the attributes of drugs and of drug users far beyond most individuals' knowledge. It is important to maintain up-to-date information in readily available forms wherever drugs are dispensed.

PROCESS GUIDE FOR DRUG–NUTRIENT INTERACTIONS IN ELDERLY CARDIAC PATIENTS

Elderly cardiac patients include those with hypertension and those with congestive heart failure. Hypertensive patients on diuretics and other antihypertensive drugs can lose mineral balance and nutrition. Patients with congestive heart failure take digitalis (digoxin), which in high and prolonged dosages can cause nausea, loss of appetite, and severe loss of weight. Diuretics in these patients can lead to serious potassium and magnesium deficiency. Potassium deficiency, the major feature of digitalis poisoning, leads to potentially fatal arrhythmia. Risk factors for potassium deficiency in cardiac patients on diuretics are low intake of foods that contain potassium and use of regular laxatives such as phenolphthalein, bisacodyl, and senna.

RISK ASSESSMENT

1   Review the medication list of hypertensives and patients with congestive heart failure for drugs contributing to the risk of potassium deficiency.
2   Examine patients for unusual muscle weakness signifying potassium deficiency.
3   Check the electrolytes for potassium deficiency. A patient is potassium deficient or hypokalemic if the serum potassium value is <3.5 mmol/liter.
4   If the patient who is potassium deficient also takes digitalis (Digoxin, Lanoxin), examine him/her and the EKG for cardiac arrhythmia indicative of digitalis poisoning.

Check for Diuretics.   Diuretics commonly taken by hypertensives are hydrochlorothiazide (Hydrodiuril, Esidrix, Dyazide), methclothiazide (Enduron), hydroflumethiazide (Salutensin), bendroflumethiazide (Naturetin), and chlorthalidone (Hygroton). Diuretics taken by patients with congestive heart failure also include furosemide (Lasix), ethacrynic acid (Edecrin), and bumetane (Bumex). Patients on any of these diuretics are at special risk of severe potassium deficiency if they are regular laxative users.

Check for Laxatives.   Laxatives that can contribute to potassium deficiency, if taken regularly by patients on diuretics, include phenolphthalein (Agoral, Evac-U-Gen, Ex-Lax, Trilax, Alophen, Feen-A-Mint, and Correctol), bisacodyl (Dulcolax), and senna (Senokot, Fletcher's Castoria, and Dr. Caldwell's Senna Laxative). *Remember:* Though laxatives may not be on the medication list, they may be taken by the patient without a doctor's order.

5   Check all patients on diuretics for sodium depletion. Severe sodium depletion can occur in patients on diuretics who are also on a strict low-sodium diet, especially in the summer, when they may also be losing sodium in sweat. Sodium depletion or hyponatremia is present when the serum sodium is <136 mmol/liter.
6   Check the body weights of all patients on digitalis at weekly intervals. Weight loss in these patients may indicate loss of edema fluid; large weight losses accompanied by loss of appetite may be indicative of digitalis poisoning.

INTERVENTION

1   Patients who are potassium deficient need more potassium or potassium supplements in their diets. Note, however, that in patients with impaired kidney function, high doses of potassium can lead to hyperkalemia and renal failure. Increasing the potassium in the diet is safer than giving supplements. Excellent food sources of potassium include bananas, orange juice, and green leafy vegetables. Don't forget to curtail laxatives.

2   Patients who are sodium deficient need to discontinue diuretics and increase their sodium intake until the serum sodium returns to normal values.

3   Patients on digitalis who are losing body weight need to have their dose reduced. Their food intake should be monitored to check that appetite returns with drug dose adjustment.

RECORDS

Identification of adverse effects of drug–nutrient interactions in cardiac patients, as well as the intervention and the response to intervention, must be recorded in the patients' charts.

PROCESS GUIDE ON ELDERLY ALCOHOLICS IN GERIATRIC CARE FACILITIES

1   Alcoholics in geriatric care facilities may be identified by their social and medical histories, by the admission diagnosis, or by their drinking practices following admission.

2   Alcoholics are at high nutritional risk, particularly if they have liver disease.

3   Alcoholics may show large unscheduled changes in body weight that are explained by their precarious eating habits, by an impaired capacity to absorb or utilize nutrients, because of muscle wasting, and because of fluid accumulation as ascites or edema.

4   Alcoholics may have special dietary requirements because they are malnourished and because of a reduced tolerance for protein.

RISK ASSESSMENT

1   For the diagnosis of alcoholism. Examine the patient's chart for a history of alcohol abuse or a diagnosis of alcoholic liver, or of pancreatic or brain disease.

2   For malnutrition. Weigh and measure the patient. Measurements in addition to weight should include mid-arm circumference. Check

TABLE 10.2 DRUG INTERACTIONS WITH ALCOHOL: RISKS OF REACTIONS WHEN THEY OCCUR AND HOW TO AVOID THEM

Drug Group	Generic Names	Brands	Risks of drinking and when not to drink
Analgesics	Acetaminophen	Tylenol	Risks of liver damage from these drugs is increased in alcoholics.
	Aspirin	Bufferin	Aspirin and alcohol both cause irritation of the stomach and taken together, stomach bleeding may occur in the alcoholic.
Antidepressants	Amitriptyline	Elavil Triavil	Motor skills are reduced when alcohol is taken with these drugs. Side effects are more common in alcoholics.
	Phenelzine	Nardil	Antidepressants like phenelzine inhibit the metabolism of amines in fermented beverages and foods. Drinking wine or beer or eating cheese when on these drugs can cause dangerous attacks of hypertension and therefore must be avoided.
Antidiabetics	Chlorpropamide	Diabenese	A single small drink can cause flushing when taken after this drug. It is therefore unwise for the patient to drink.
Anticonvulsants	Phenytoin	Dilantin	Seizures and alcohol can cause loss of consciousness. People on drugs for control of seizures should not drink alcohol.

the hematology lab report for anemia and high MCV values, indicative of folate deficiency. Check the chemical profile for high transaminase values (SGOT), indicative of liver disease, and low serum albumin levels indicative of protein deficiency.

3   For recognition of potential adverse drug and drug–nutrient reactions. Check the patient's medication list for drugs that are not tolerated by alcoholics, including those that produce sedation or interfere with liver function. Check that the patient is not taking a drug with which an unpleasant reaction will occur if alcohol is consumed; an example is chlorpropamide (Diabenese). Check if the patient is buying or being given alcoholic beverages.

INTERVENTION

1   With alcohol abuse: Ensure that the patient does not have access to alcohol.
2   With malnutrition: Provide nutritional support with supplements and select a formula that can be tolerated by the patient.
3   With risk of adverse drug reactions: Stop all drugs that are dangerous for the alcoholic, including over-the-counter drugs containing alcohol.

RECORDS

Be sure that the patient's chart identifies the patient as alcoholic.

FOLLOW-UP

1   Watch that the patient does not regain access to alcohol.
2   Check that the patient's nutritional status improves with nutritional support.
3   Check that no new drugs are prescribed that could be harmful to the patient.

PROCESS GUIDE FOR DRUG–NUTRIENT INTERACTIONS IN ELDERLY ARTHRITIC PATIENTS

Elderly arthritic patients include those with osteoarthritis, gout, and rheumatoid arthritis. Patients who have osteoarthritis take aspirin or other anti-inflammatory drugs to relieve joint pain. These drugs can cause bleeding into the intestine, leading to anemia. Ibuprofen is also taken by these patients, and it can cause fluid retention leading to weight gain. Patients with gout usually take colchicine during acute gout attacks. Colchicine can impair nutrient absorption and can cause weight loss. Patients with rheumatoid arthritis who are frequently debilitated take cortisone-like drugs or penicillamine in order to control their disease, to reduce joint swelling, or to prevent deformity. Cortisone-like drugs can increase body weight, elevate blood sugar, and also impair calcium absorption and worsen osteoporosis. Penicillamine can cause zinc deficiency, which leads to loss of appetite, impaired wound healing, and development of a rash. This drug can also cause vitamin B_6 deficiency, which is associated with anemia.

RISK ASSESSMENT

Review the medication list of patients with arthritis for drugs contributing to the risk of anemia, weight loss or gain, high blood sugar, or severe osteoporosis with risk of fractures. Examine the lab data for changes in the blood count, including reduction in hemoglobin and hematocrit, as well as for changes in blood sugar. Find out whether the patient has gained or lost weight.

Common anti-arthritic drugs that contribute to anemia are aspirin (including Bufferin and other buffered brands), indomethacin (Indocin), naproxen (Naprosyn) and penicillamine (Cuprimine or Depen). Common anti-arthritic drugs that may cause weight gain include the cortisone-like group such as prednisone (Deltasone), methylprednisolone (Medrol) or triamcinolone (Aristocort), and ibuprofen (Motrin, Rufen, or Nuprin). These drugs can also cause loss of control of diabetes and can, over time, increase the severity of osteoporosis. Drugs causing weight loss include penicillamine and colchicine.

INTERVENTION (by the M.D. or physician's assistant)

For anemia: Discontinue aspirin, indomethacin, naproxen, or penicillamine. For weight gain: Reduce dose of cortisone-like drugs or stop ibuprofen. For weight loss: Stop penicillamine or colchicine. For elevated blood sugar: Stop cortisone-like drugs, particularly in diabetics.

RECORDS

Identify all potential drug–nutrient interactions related to anti-arthritic drugs on the patient's medication list and record action taken. If adverse effects have already occurred, these should be listed in the progress notes. All notations related to drug–nutrient interactions and anti-arthritic drugs should be identified with a blue star.

FOLLOW-UP

Check patients showing adverse effects related to anti-arthritic drugs for improvement in anemia, return to normal body weight, and normalization of blood sugar. For patients with drug-related anemia due to blood loss, an iron supplement will be required.

PROCESS GUIDE ON DRUG–NUTRIENT INTERACTIONS FOR ELDERLY PATIENTS WITH NEUROLOGICAL OR MENTAL HEALTH PROBLEMS

1   Tranquilizers such as chlorpromazine increase the patient's appetite and a patient with free access to food may experience weight gain. A similar increase in appetite and weight can occur with lithium carbonate.
2   Lithium carbonate blood levels and toxicity are increased by change to a low-sodium diet while the patient is on the drug.
3   Anticonvulsant drugs such as phenytoin and phenobarbital can induce nutritional deficiencies. The patient may develop vitamin D deficiency, with loss of calcium from the bones, or folate deficiency leading to anemia. Those taking these anticonvulsant drugs who do not go outdoors or who drink no milk are at highest risk of vitamin D deficiency.

IDENTIFICATION OF RISKS

1   Check the medical records for unscheduled weight gain of patients on phenothiazine tranquilizers including Thorazine, Largactil, Mellaril, Navane, and lithium preparations such as Eskalith or Cibalith-S.
2   Check the medical records of patients on anticonvulsant drugs such as Dilantin (phenytoin) or phenobarbital for low blood calcium or phosphorus levels as well as for low hemoglobin or red cell count.
3   Check the medical records of patients on lithium for new low-sodium diet orders. Note also that sodium depletion and increased risk of lithium toxicity can occur if these patients are given diuretics.

INTERVENTION

1   Weigh and chart the weights of patients who are on phenothiazine tranquilizers and lithium. If weight gain occurs, report this to the MD or PA and to the dietary department. Patients showing weight gain should have their caloric intake reduced by diet prescription and should not have access to snack foods.
2   Check blood levels of patients on lithium, especially if there is a change in the diet or medication prescription with risk of sodium depletion. Be vigilant that these patients not be given strict low-sodium diets unless their drug dosage is appropriately reduced.
3   For patients on anticonvulsant drugs, if evidence is obtained that

vitamin D deficiency is a risk, give a vitamin D supplement; an appropriate dose is 1000 IU/day. If the patient on an anticonvulsant drug is becoming anemic, as shown by a falling hemoglobin or red cell count, check plasma folate values. If plasma folate is <3 ng/ml, give a folic acid supplement of 1 mg/day. Do not give a higher dose of folic acid—it can interfere with the absorption of Dilantin.

RECORDING AND REPORTING

When the patient's chart shows any of the following, place a blue star on the patient order sheet and record the problem (always report the problem to the MD or PA and the dietitian):

1   Weight gain >5 lb. in one month in patients on phenothiazine tranquilizers or lithium; or
2   Risk of lithium toxicity (low-sodium diet or diuretic) in a patient on lithium carbonate; or
3   Evidence of anemia (decreasing hemoglobin or red cell count) or of vitamin D deficiency (low serum calcium or phosphorus) in a patient on Dilantin or phenobarbital.

FOLLOW-UP

1   For weight gain in patients on phenothiazine tranquilizers and lithium: Check that weight gain stops with change of diet.
2   For patients on lithium given low-sodium diets or a prescription for a diuretic, check that the order is canceled; if the diet or drug is essential, the dose of lithium should be reduced.
3   For patients on Dilantin and/or phenobarbital, check that the folic acid supplement corrects anemia and that the vitamin D supplement corrects low serum calcium and phosphorus levels.

PROCESS GUIDE FOR NURSES ON DRUG–NUTRIENT INTERACTIONS IN DIABETICS

1   *What are you looking for?*
    a.   Drugs that are increasing the patient's blood sugar, or
    b.   Drugs that are decreasing the patient's blood sugar.
2   *Check the medication list in the patient's chart:*

a. For corticosteroids, thiazides, and cold or cough medicines that contain ephedrine-like drugs. All these medications can elevate blood sugar.

b. For concurrent prescription of oral antidiabetic drugs and aspirin and/or propranolol (Inderal). Also find out if the patient has been drinking alcohol concurrent with insulin or oral antidiabetic drugs. All these drug combinations can cause hypoglycemia, which can be dangerous to the patient.

3 *Look at the blood chemistry reports.* If the blood glucose has increased since introduction of the hyperglycemic drug in question, this may indicate a causal relationship. Similarly, if the patient's blood glucose has decreased unexpectedly with or without concurrent hypoglycemic attacks, this suggests that a drug (other than the drug which the patient has had prescribed for control of his/her diabetes) is inducing this response. Don't forget that this drug may be alcohol!

4 *What action do you take?*
Identify the drug that you suspect is affecting the patient's diabetes. If you have evidence that the patient is drinking alcohol, make a note to this effect on the medication list and star the notation. Report your findings to the physician or physician's assistant (PA) who is responsible for the medical care of the patient.

5 *Check that the MD or PA stops the drug in question and initializes the medication list.*

6 *How should you follow-up?*
Be sure that the patient has weekly blood glucose determinations until the diabetes is controlled.

Chapter 11

PRESCRIPTION GUIDELINES

In order to assist health care providers and patients in identifying risks of drug and diet interactions, this chapter discusses information on prescription and nonprescription drugs. Generic and brand names; indications for use; common dosage regimens; food, nutrient, alcohol interactions; and drug timing are listed in Tables 11.1 through 11.16. Key references are provided in Table 11.17. The material is presented in two forms to provide instruction for health professionals and patients, respectively.

IDENTIFICATION OF DRUG–NUTRIENT INTERACTIONS: A GUIDE FOR HEALTH PROFESSIONALS

Drug–nutrient interactions include (1) physiochemical interactions in the GI tract that reduce the absorption of drugs and/or nutrients; (2) metabolic interactions after absorption that impair drug or nutrient utilization; and (3) function interactions that alter the rate of elimination of drugs or nutrients.

Some adverse results of drug–nutrient interactions are: (1) failure of nutritional support; (2) nutritional deficiency; (3) unwanted weight change; (4) loss of disease control; and (5) toxic reactions.

There is potential for a drug–nutrient interaction in a patient if:

1 The patient is receiving drugs at mealtimes.
2 The patient on drugs has a major change in the protein content of his/her diet.
3 The patient is given a high-potency vitamin or mineral supplement.
4 The patient is receiving drugs via a nasogastric or other enteral feeding tube.
5 The drugs being administered cause malabsorption or have an antinutrient effect.

TABLE 11.1 DRUG–NUTRIENT INTERACTION GUIDE

Generic Drug	When to Give Drug	Diet Risk	Weight Change	Nutritional Risk	Metabolic Effect
Acetaminophen	OK w/food	None	None	None	None
Acetazolamide	With food	Lo K$^+$	Loss	K$^+$ def	Hypokalemia
Allopurinol	With food	Hi purine	None	None	None
Aluminum hydroxide	2 h. pc	Low Ca^{++}	None	Phosphate def	Hypophosphatemia
Amantadine	With food	None	None	None	None
Amiloride	OK w/food	Salt subst.	Loss	None	Hyperkalemia
Amitriptyline	With food	Hi kcal	Gain	Obesity	SIADH
Amoxacillin	OK w/food	None	None	None	None
Ampicillin	OK w/food	None	None	None	None
Aspirin	With food	None	None	Anemia	Hypoglycemia (D)
Atenolol	With food	None	None	None	Hypoglycemia (D)
Baclofen	With food	None	None	None	Hyperglycemia (D)
Benzotropine	OK w/food	None	None	None	None
Bethanechol	1 H. ac	None	None	None	None
Bisacodyl	OK w/food	Low K$^+$	Loss	K$^+$ def	Hypokalemia
Bromocriptine	OK w/food	None	None	None	None
Bumetanide	OK w/food	Hi Na$^+$	Loss	K$^+$ def	Hypokalemia
Captopril	Fasting	K$^+$ suppl.	None	None	None
Carbamezepine	With food	None	None	None	SGOT elevated
Cefaclor	1 h. ac	None	None	Vit. K def	None
Cephalexin	1 h. ac	None	None	None	None
Chlorpheniramine	OK w/food	None	None	None	Hyperglycemia
Chlorpromazine	OK w/food	Hi kcal	Gain	Obesity	SGOT elevated
Chlorpropamide	30 min. ac	Hi kcal	None	None	Hypoglycemia
Chlorthalidone	With food	Low Na$^+$	Loss	K$^+$ & Na$^+$ def	Hyperglycemia

Cholestyramine	With juice	Hi fat	Loss	Malabsorption	Hypocalcemia
Cimetidine	30 min. ac	Caffeine	None	Vit. B_{12} def	None
Colchicine	OK w/food	Hi lactose	Loss	Malabsorption	SGOT elevated
Colestipol	With juice	Hi fat	Loss	Malabsorption	Hypocalcemia
Conjugated estrogens	OK w/food	None	Gain	None	None
Cyproheptadine	OK w/food	Hi kcal	Gain	Obesity	None
Desipramine	With food	Hi kcal	Gain	Obesity	SIADH
Digitalis	1 h. ac	Hi fiber	Loss	Cachexia	Tox./hypokalemia
Diltiazem	1 h. ac	None	None	None	None
Diphenhydramine	OK w/food	None	None	None	None
Dipyridamole	2 h. ac	None	None	None	None
Docusate sodium	Bedtime	None	None	None	None
Doxepin	With food	Hi kcal	Gain	Obesity	Hyperglycemia
Ephedrine	OK w/food	Coffee	Loss	None	Hyperglycemia
Erythromycin	1 h. ac	None	None	None	None
Ethacrynic acid	With food	Lo K^+	Loss	K^+ def	Hypokalemia
Fenoprofen	With food	None	Rare	Anemia	Hyperkalemia
Fluphenazine	OK w/food	Hi kcal	Gain	Obesity	SGOT elevated
Folic acid	With food	None	None	None	None
Furosemide	1 h. ac	Low K^+	Loss	Dx B_{12} def	Hypokalemia
Haloperidol	OK w/food	None	None	K^+ & Mg def	None
Hydralazine	With food	None	None	None	SGOT elevated
Hydrochlorothiazide	With food	Lo Na^+	Loss	Vit B_6 def	Hyperglycemia
Hydroxyzine	OK w/food	Hi kcal	Gain	K^+ & Na def	None
Ibuprofen	With food	None	Gain	Obesity	Hyperkalemia
Imipramine	With food	Hi kcal	Gain	Anemia	SIADH
Indomethacin	With food	None	None	Obesity	Hyperkalemia
Isocarboxazid	OK w/food	Caffeine/His, Tyr	None	Anemia	SGOT elevated
Isoniazid	Fasting	Lo vit. B_6/His, Tyr	None	B_6 & D def	SGOT elevated

(continued)

TABLE 11.1 DRUG–NUTRIENT INTERACTION GUIDE (CONT.)

Generic Drug	When to Give Drug	Diet Risk	Weight Change	Nutritional Risk	Metabolic Effect
Isorbide dinitrate	Fasting	None	None	None	None
Levodopa + carbidopa	OK w/food	None	None	None	None
Lithium carbonate	With food	Lo Na$^+$ & Hi kcal	Gain	Obesity	T4 depressed
Meclizine	With food	None	None	None	None
Methotrexate	Fasting	None	Loss	Folate def	SGOT elevated
Methyldopa	2 h. pc	Hi Na$^+$	Gain	Nausea	Hyperkalemia
Metoclopropamide	30 min. ac	None	None	None	None
Metolazone	With food	Lo K$^+$	Loss	K$^+$ def	Hypokalemia
Metoprolol	With food	None	None	None	None
Molindone	OK w/food	Hi kcal	Gain	Obesity	SGOT elevated
Nalbuphine	OK w/food	None	Loss	Anorexia	None
Nalidixic acid	With food	None	None	None	SGOT elevated
Nifedipine	With food	None	None	None	Hypoglycemia
Nitrofurantoin	With food	None	None	None	None
Nitroglycerin	OK w/food	None	None	None	Methemoglobin
Nortriptyline	With food	Hi kcal	Gain	Obesity	SIADH
Pargyline	OK w/food	Caffeine	None	None	SGOT elevated
Penicillamine	2 h. ac	Lo fluid	Loss	B$_6$ & Zn def	Hypozincemia
Penicillin V	Fasting	None	None	None	None

Drug					
Pentoxifylline	With food	Hi caffeine	None	None	None
Phenazopyridine	With food	None	None	None	Methemoglobin
Phenelzine	OK w/food	Caffeine/His, Tyr	None	vit. B$_6$ def	SGOT elevated
Phenoxybenzamine	OK w/food	None	None	Anorexia	None
Phenylephrine	OK w/food	None	None	None	Hyperglycemia
Phenylpropanolamine	OK w/food	None	Loss	None	Hyperglycemia
Phenytoin	1 h. ac	Lo folate	None	Folate & D def	None
Piroxicam	With food	None	Gain	Anemia	Hyperkalemia
Potassium supp.	With food	Hi K$^+$	None	Vit B$_{12}$ def	Hyperkalemia
Prazosin	OK w/food	Hi Na$^+$	Gain	Na$^+$ retention	Hypernatremia
Prednisone	With food	Hi Na$^+$	Gain	Osteoporosis	Hyperglycemia
Primidone	OK w/food	Lo folate	None	Folate def	None
Procainamide	1 h. ac	None	None	None	SGOT elevated
Procarbazine	With food	Caffeine/His, Tyr	Loss	None	SGOT elevated
Prochlorperazine	OK w/food	None w/Rx	None	None w/Rx	SGOT elevated
Propranolol	With food	None	None	None	Hypoglycemia (D)
Pseudoephedrine	OK w/food	None	None	None	Hyperglycemia
Pyridostigmine	OK w/food	None	None	None	None
Quinidine	With food	Caffeine	Loss	None	None

(continued)

195

TABLE 11.1 DRUG–NUTRIENT INTERACTION GUIDE (CONT.)

Generic Drug	When to Give Drug	Diet Risk	Weight Change	Nutritional Risk	Metabolic Effect
Ranitidine	30 min. ac	Caffeine	None	Vit B_{12} def	None
Spironolactone	With food	Salt subst.	Gain	None	Hyperkalemia
Sucralfate	With food	None	None	None	None
Sulfasalazine	2 h. ac	Lo folate	None	Folate def	Hypoglycemia
Sulindac	With food	None	Gain	Anemia	Hypoglycemia (D)
Terbutaline	OK w/food	Hi coffee	None	Anorexia	Hyperglycemia
Tetracycline	2 h. ac	Milk based	None	Ca^+ loss	None
Theophylline	1 h. ac	Hi protein	None	None	None
Thioridazine	OK w/food	Hi kcal	Gain	Obesity	SGOT elevated
Thiothixene	OK w/food	Hi kcal	Gain	Obesity	SGOT elevated
Thyroxine	30 min. ac	Low kcal	Loss	None	BMR elevated
Tolazamide	30 min. ac	Hi kcal	None	None	Hypoglycemia
Tranylcypromine	OK w/food	Caffeine/ His, Tyr	None	None	SGOT elevated
Trazodone	OK w/food	Caffeine	None	None	None
Triamterene	Fasting	None	None	Folate def	Hyperkalemia
Trihexyphenidyl	OK w/food	Lo fiber	None	Constipation	None
Trimethoprim/sulfa	OK w/food	Lo fluid	None	Folate def	None
Verapramil	Fasting	None	None	Constipation	None
Warfarin	2 h. pc	Hi vit K	None	Vit K def	Hypoprothrombin

TABLE 11.2 DRUGS THAT CAUSE ANEMIA

Due to blood loss
Aspirin (Easprin, Bufferin)
Fenoprofen (Nalfon)
Ibuprofen (Motrin)
Indomethacin (Indocin)
Piroxicam (Feldene)
Sulindac (Clinoril)

Due to failure of cell maturation
Methotrexate (Mexate)
Phenytoin (Dilantin)
Primidone (Mysoline)
Triamterene (Dyrrenium)
Trimethoprim (Bactrim)

TABLE 11.3 DRUGS THAT AFFECT CALCIUM AND PHOSPHATE STATUS

Drugs that cause hypocalcemia
Cholestyramine (Questran)
Colestipol (Colestid)
Ethacrynic acid (Edecrin)
Furosemide (Lasix)
Isoniazid (INH)
Phenytoin (Dilantin)
Primidone (Mysoline)

Drugs that cause hypercalcemia
Thiazides, examples include:
  Bendroflumethiazide (Naturetin)
  Hydrochlorothiazide (Hydrodiuril; Esidrix)
  Methylclothiazide (Enduron)
Vitamin D and metabolites, examples include:
  Calcifediol (Calderol)
  Calcitriol (Rocaltrol)
  Ergocalciferol (Calciferol)

Drugs that cause hypophosphatemia
Aluminum-containing antacids, examples include:
  Aluminum hydroxide gels (Amphogel)
  Alumina and magnesia tabs (Maalox)
Phenytoin (Dilantin)

TABLE 11.4 DRUGS THAT CAUSE ELECTROLYTE IMBALANCE

Drugs that cause hyponatremia
Amitriptyline (Elavil)
Captopril (Capoten)
Chlorpropamide (Diabenese)
Spironolactone (Aldactone)
Thiazides, examples include:
    Bendroflumethiazide (Naturetin)
    Hydrochlorothiazide (Hydrodiuril; Esidrix)
    Methylclothiazide (Enduron)
Thioridazine (Mellaril)

Drugs that cause hypokalemia
Acetazolamide (Diamox)
Bisacodyl (Dulcolax)
Bumetanide (Bumex)
Corticosteroids, examples include:
    Hydrocortisone (Cortef)
    Methylpredisolone (Medrol)
    Prednisone (Deltasone)
    Prednisolone (Delta-Cortef)
Ethacrynic acid (Edecrin)
Furosemide (Lasix)
Lithium carbonate (Eskalith)
Metolazone (Zaroxolyn)
Salicylates, including aspirin (in high doses)
Thiazides, examples include:
    Bendroflumethiazide (Naturetin)
    Hydrochlorothiazide (Hydrodiuril; Esidrix)
    Methylclothiazide (Enduron)

Drugs that cause hyperkalemia
Amiloride (MIdamor)
Indomethacin (Indocin)
Methyldopa (Aldomet)
Other NSAIDs, examples include:
    Fenoprofen (Nalfon)
    Ibuprofen (Motrin)
    Piroxicam (Feldene)
    Sulindac (Clinoril)
Isoniazid
Spironolactone (Aldactone)
Triamterene (Dyrrenium)

TABLE 11.5 MONAMINE OXIDASE INHIBITOR INCOMPATIBILITIES

Monamine oxidase inhibitor drugs when taken with high-tyramine foods cause attacks of hypertension and can precipitate stroke.

High tyramine foods
  Cheese
  Meat extracts
  Yeast extracts
  Sauerkraut
  Dry sausage
  Smoked or pickled fish
  Livers
  Beer and ale
  Red and white wines

Monamine oxidase inhibitor drugs when taken with high histamine foods cause flushing, itching, and headaches.

High histamine foods
  Tuna and bonito

Monamine oxidase inhibitor drugs

Generic	Brand Names
Isocarboxazid	Marplan
Isoniazid	Nydrazid
	INH
Pargyline	Eutonyl
Phenelzine	Nardil
Procarbazine	Matulane
	Natulan
Tranylcypromine	Parnate

TABLE 11.6 SPECIFIC ALCOHOL–DRUG REACTIONS

Drugs that cause sedation and result in frequent falls if alcohol is taken with them
  Barbiturates
  Tranquilizers

Drug that causes loss of motor skills and results in frequent falls if alcohol is taken
  Antidepressants

Drug that may cause flush in diabetics if alcohol is taken
  Chlorpropamide

Drugs that cause liver damage in heavy alcohol consumers
  Acetaminophen
  Methotrexate

TABLE 11.7 METABOLIC EFFECT OF DRUGS

Drugs that cause hyperglycemia (in diabetics)
Baclofen (Lioresal)
Corticosteroids, examples include:
   Hydrocortisone (Cortef)
   Methylpredisolone (Medrol)
   Prednisone (Deltasone)
   Prednisolone (Delta-Cortef)
Ephedrine
Phenylephrine
Phenylpropanolamine
Pseudoephedrine
Thiazides, examples include:
   Bendroflumethiazide (Naturetin)
   Hydrochlorothiazide (Hydrodiuril; Esidrix)
   Methylclothiazide (Enduron)
Drugs that cause hypoglycemia (in diabetics)
Alcohol
Aspirin
Atenolol (Tenormin)
Insulin
Propranolol (Inderal)
Sulfa drugs
Sulfonylurea drugs, examples include:
   Acetohexamide (Dymelor)
   Chlorpropamide (Diabinese)
   Glipizide (Glucotrol)
   Glyburide (Diabeta; Micronase)
   Tolbutamide (Orinase)
Sulindac (Clinoril)

6 The patient is drinking alcoholic beverages or is otherwise nutritionally compromised.
7 The patient is on long-term treatment with one or more drugs that affect nutritional status.
8 The patient is on a drug that significantly affects appetite.

The risk of drug–nutrient interactions in patients can be identified if you do the following:

1 Check the medication list for drugs that impair nutritional status and/or affect appetite, and for nutrient supplements that interfere with the intended drug effects.
2 Check the relationship between drug times and mealtimes.

TABLE 11.8 FORMULA INCOMPATIBILITIES

Drugs that are incompatible with formulas
   Cibalith-S syrup
   Dilantin suspension
   Dimetane elixir
   Dimetapp elixir
   Feosol elixir
   KCL liquid
   Klorvess syrup
   Lanoxin elixir
   Mandelamine suspension
   Mellaril oral
   Neo-Calglucon syrup
   Robitussin expectorant
   Sudafed syrup
   Thorazine concentrate

(Drugs listed here are acidic liquid formulations that are incompatible with enteral feeding formulas. These drugs should not be given in tube-fed patients without first stopping the flow of formula and irrigating the tube with water. Other drugs besides those listed here may block the tube if given as solid preparations. Additionally, broad-spectrum antibiotics often cause diarrhea in tube-fed patients.)

3 Consider the types and formulations of drugs being given via feeding tubes.
4 Evaluate whether changes in diet orders could affect the duration of a drug's action.
5 Check progress notes for evidence of loss of disease control since a drug was prescribed or a diet was changed.
6 Check and monitor lab reports for evidence of nutrient depletion or nutritional anemia following prescription of a drug.
7 Consider relationships between drug intake and weight change.
8 Check the patient's history for evidence of alcohol abuse.

USE OF THE GUIDE

Drug Lists. In order to see whether patients are at risk for specific types of drug–nutrient interactions, information should be sought from Table 11.1. Table 11.1 gives information on when the drug should be given in relation to food, the risks of drug–food interactions, weight changes likely, nutritional deficiencies, and adverse metabolic effects.

Reference Tables. Reference tables list drugs causing particular nutritional deficiencies, electrolyte imbalances, and metabolic effects. Table 11.2 shows drugs causing anemia; Table 11.3 shows drugs affecting

TABLE 11.9 MEDICATIONS THAT CONTAIN SUGAR

Cough syrups
  Coricidin Cough Syrup (Schering)
  Triaminicol Decongestant Cough Syrup (Dorsey)
  Vicks Cough Syrup (Vicks Health Care)
Expectorants
  Cheracol D Cough Syrup (Upjohn)
  Formula 44D Decongestant Cough Mixture (Vicks Health Care)
  Guaifenesin Syrup (Lederle)
Throat lozenges
  Vicks Throat Lozenges
Allergy relief products
  Chlor-Trimeton Allergy Syrup (Schering)
  Pseudoephedrine Syrup (Comatic)
  Rhinosyn-X Syrup (Comatic)
Analgesics
  Liquiprin (Norcliff Thayer)
  Tylenol Extra Strength Acetaminophen Adult Liquid Pain Reliever
    (McNeil Consumer Products)
Antibiotics
  EES 400 Liquid (Abbott)
  Sumycin Syrup (Squibb)
Tranquilizers (antipsychotic drugs)
  Mellaril Suspension (Sandoz)
  Thorazine Syrup (Smith Kline and French)

(These medications may provide a significant calorie source.)

calcium and phosphorus; and Table 11.4 shows drugs causing sodium and potassium imbalances.

Table 11.5 shows incompatibilities of amine-containing foods with monamine oxidase inhibitor drugs. Table 11.6 shows drug–alcohol interactions; Table 11.7 lists drugs causing hyper- or hypoglycemia in diabetics, and Table 11.8 shows drugs incompatible with formula foods.

Further tables show drugs containing nutrients or alcohol that may contribute significantly to daily intake. Table 11.9 lists drugs containing sugar; Table 11.10 shows drugs containing alcohol; Table 11.11

TABLE 11.10 MEDICATIONS THAT CONTAIN ALCOHOL

	% Alcohol
*Cough medicines**	
Brondecon Elixir (Parke-Davis)	20
Hills Decongestant Cough Formula Liquid	22
Isuprel Compound Elixir (Breon)	19
Lufyllin Elixir (Wallace)	17
Mudrane-GG Elixir (Polythress)	20
Neothylline-GG Elixir (Lemmon)	10
NyQuil Nighttime Cold Medicine Liquid (Vicks)	25
Antidiarrheal	
Parelixir (Purdue-Frederick)	18

(See Table 11.6 for specific Alcohol–Drug Interactions.)

*Cough medicines designated as elixirs contain alcohol unless stated to be alcohol-free.

shows drugs containing potassium; and Table 11.12 shows drugs containing sodium. Abbreviations used in the tables are shown in Table 11.13.

GUIDE TO PATIENTS ON HOW TO REDUCE THE RISKS OF INTERACTIONS OF MEDICINES WITH FOODS

Safe use of medications requires that you know the generic or brand names of the medications you take. The generic and brand names of common drugs can be found in Table 11.14, together with the names of the manufacturers. Alternate brands will be found in Table 11.15.

TABLE 11.11 POTASSIUM CONTENT OF SOME MEDICATIONS

Mysteclin F	21	mg/5 ml
Neutra-Phos	278	mg/capsule
Neutra-Phos-K	556	mg/capsule
Penicillin G Potassium	66.3	mg/million units
Pfizerpen VK	1.7	mg/250 mg

TABLE 11.12 SODIUM CONTENT OF SOME MEDICATIONS

Oral

Alka-Seltzer, antacid	296	mg/tab
Alka-Seltzer, pain reliever	551	mg/tab
Alka-Seltzer plus	482	mg/tab
Alevaire	80	mg/5 ml
Di-Gel	10.6	mg/tab
Di-Gel, liquid	8.5	mg/5 ml
Dristan Cough Formula	58	mg/5 ml
Fleet Phospho-Soda	550	mg/5 ml
Phosphalgel	12.5	mg/5 ml
Rolaids	53	mg/tab
Sodium Salicylate, tab	49	mg/5 gr
Titralac liquid	11	mg/5 ml
Vicks Cough Syrup	41	mg/5 ml
Formula 44D Decongestant Cough Mixture	51	mg/5 ml

Other

Neutra-Phos	164	mg/capsule

Parenteral

Azlocillin	49.9	mg/gm
Cefoxitin	52.9	mg/gm
Mezlocillin	42.6	mg/gm
Pen G Potassium	12.2	mg/gm
Pen G Sodium	62.6	mg/gm
Piperacillin	45.5	mg/gm
Ticarcillin	119.6–149.5	mg/gm
Saline solutions		
0.9% NaCl	3542	mg/L
D5 0.2% NaCl	782	mg/L
D5 0.45% NaCl	1771	mg/L
D5 0.9% NaCl	3542	mg/L
Ringers	3392.5	mg/L
Ringers Lactate	2990	mg/L

TABLE 11.13 KEY TO ABBREVIATIONS

Abbreviation	Meaning
Ca^{++}	Calcium
(D)	In diabetics
Fasting	Drug should be given 1–2 hours before food.

TABLE 11.13 KEY TO ABBREVIATIONS (CONT.)

Abbreviation	Meaning
1 h. ac	1 hour before food
1 h. pc	1 hour after food
His	Histamine containing foods
K^+	Potassium
Lo	Low
Mg	Magnesium
Na^+	Sodium
NSAID	Nonsteroidal anti-inflammatory drug
OK w/food	Drug can be given with food
Salt subst	Salt substitute
SGOT	Serum glutamic oxalacetic transaminase
SIADH	Syndrome of inappropriate antidiuretic hormone secretion
T_4	Thyroxine
Tyr	Tyramine-containing foods

TABLE 11.14 GENERIC AND COMMON BRAND NAMES (CONT.)

Generic Name	Brand Name	Manufacturer
Aluminum hydroxide gel	Amphojel	Wyeth
Amiloride HCL	Midamor	MSD[1]
Aspirin	Bufferin	Bristol-Myers
Aspirin-containing antacid	Alka-Seltzer	Miles Labs
Atenolol	Tenormin	Stuart
Carbidopa-Levodopa	Sinemet	MSD
Chlorpromazine HCL	Thorazine	SKF[2]
Chlorpropamide	Diabenese	Pfizer
Cholestyramine	Questran	Mead Johnson
Cimetidine	Tagamet	SKF
Clonidine HCL	Catapres	Boehringer Ingelheim
Colchicine	Colchicine	Abbott
Diazepam	Valium	Roche
Digoxin	Lanoxin	Burroughs-Wellcome
Furosemide	Lasix	Hoechst-Roussel
Griseofulvin	Fulvicin P/G	Schering
Guanethidine sulfate	Ismelin sulfate	Ciba
Hydralazine	Apresoline	Ciba
Hydrochlorothiazide	Esidrix	Ciba
Isoniazid (INH)	Nydrazid	Squibb
Isotretinoin [13]Cisretinoic acid	Accutane	Roche
Lithium carbonate or citrate	Eskalith	SKF
Methotrexate	Methotrexate	Lederle

TABLE 11.14 GENERIC AND COMMON BRAND NAMES (CONT.)

Generic Name	Brand Name	Manufacturer
Methyldopa	Aldomet	MSD
Mineral oil	Petrolagar plain	Wyeth
Moxalactam disodium	Moxam	Lilly
Neomycin sulfate	Mycifradin sulfate	Upjohn
Nitrofurantoin	Furadantin	Norwich Eaton
Nitroglycerine	Nitrostat	Parke-Davis
Norethindrone Ethinylestradiol	Ortho-Novum 1/35–21	Ortho Pharm.
Norethindrone Mestranol	Ortho-Novum 1/50–21	Ortho Pharm.
Penicillamine	Cuprimine	MSD
Phenelzine	Nardil	Parke-Davis
Phenolphthalein	Alophen pills	Parke-Davis
Phenytoin	Dilantin sodium	Parke-Davis
Potassium chloride liquid	Klorvess syrup	Dorsey
Prazosin HCL	Minipress	Pfizer
Prednisone	Deltasone	Upjohn
Propranolol	Inderal	Ayerst
Ranitidine	Zantac	Glaxo
Reserpine	Serpasil	Ciba
Spironolactone	Aldactone	Searle
Sulfasalazine	Azulfidine	Pharmacia
Tetracycline	Sumycin	Squibb
Theophylline	Theo-dur	Key
Triamterene	Dyrenium	SKF
Trimethoprim/Sulfamethoxazole	Bactrim	Roche
Valproic acid	Depakene	Abbott
Warfarin sodium crystalline	Coumadin sodium	Endo

[1]MSD = Merck, Sharp and Dohme
[2]SKF = Smith Kline and French

Table 11.16 shows health problems for which particular drugs are prescribed.

You should never drink alcoholic beverages while on medications, unless your physician or pharmacist feels that it is safe to do so. Drinking these beverages while taking drugs can cause serious side effects as well as merely unpleasant reactions.

Do not change your diet while you are taking medications unless you have discussed the changes with your physician. Do not start tak-

TABLE 11.15 ALTERNATE BRANDS IN COMMON USAGE

Generic Name	Brand Name	Alternate Brands
Aluminum hydroxide gel	Amphojel	Dialume, Alu-Cap
Amiloride HCL	Midamor	Alternagel
Aspirin	Bufferin	None
Aspirin-containing antacid	Alka-Seltzer	Ecotrin, Ascriptin, A.S.A.
Atenolol	Tenormin	None
Carbidopa-Levodopa	Sinemet	None
Chlorpromazine HCL	Thorazine	Promapar. Chlorpromazine HCL
Chlorpropamide	Diabenese	Chlorpropamide
Cholestyramine	Questran	Colestid (Upjohn, similar resin)
Cimetidine	Tagamet	None
Clonidine HCL	Catapres	None
Colchicine	Colchicine	Colsalide
Diazepam	Valium	Valrelease
Digoxin	Lanoxin	SK-Digoxin
Furosemide	Lasix	SK-Furosemide
Griseofulvin	Fulvicin P/G	Gris-Peg, Grisactin Ultra
Guanethidine sulfate	Ismelin sulfate	None
Hydralazine	Apresoline	Dralzine, Hydralazine
Hydrochlorothiazide	Esidrix	Hydrodiuril, Oretic, Thiuretic
Isoniazid (INH)	Nydrazid	Isoniazid, Niconyl, Laniazid
Isotretinoin ^{13}Cisretinoic acid	Accutane	None
Lithium carbonate or citrate	Eskalith	Lithobid, Cibalith-S, Lithane
Methotrexate	Methotrexate	Mexate
Methyldopa	Aldomet	Aldoril
Mineral oil	Petrolagar plain	Agoral Plain
Moxalactam disodium	Moxam	None
Neomycin sulfate	Mycifradin sulfate	Neobiotic
Nitrofurantoin	Furadantin	Furan, Furalan
Nitroglycerine	Nitrostat	Nitroglycerin, Nitrospan
Norethindrone Ethinylestradiol	Ortho-Novum 1/35–21	Ortho-Novum 1/35–28

(continued)

ing vitamins, minerals, or over-the-counter drugs while taking prescription drugs because these can interfere with the effectiveness of the prescription drugs and can also produce side effects affecting your nutrition.

Always tell your physician about all the medications you are taking so that you can avoid harmful interactions.

Be sure that you understand how much of a drug you are to take, how often you are to take the drug, and when you are to take it in relation to food. If you do not understand, check all these points with your physician and your pharmacist.

Table 11.17 lists key references.

TABLE 11.15 ALTERNATE BRANDS IN COMMON USAGE (CONT.)

Generic Name	Brand Name	Alternate Brands
Norethindrone Mestranol	Ortho-Novum 1/50–21	Ortho-Novum 1/50–28
Penicillamine	Cuprimine	Depen Timetabs
Phenelzine	Nardil	None
Phenolphthalein	Alophen pills	Ex-Lax, Feen-A-Mint
Phenytoin	Dilantin sodium	Diphenylan, Phenytoin Sodium
Potassium chloride liquid	Klorvess syrup	KCL Liquid
Prazosin HCL	Minipress	None
Prednisone	Deltasone	Orasone, Prednisone
Propranolol	Inderal	Inderide (Inderal + HCTZ)
Ranitidine	Zantac	None
Reserpine	Serpasil	Sandril, Serpate, Zepine
Spironolactone	Aldactone	Spironolactone, Spiractone
Sulfasalazine	Azulfidine	S.A.S.-500, Sulfasalazine
Tetracycline	Sumycin	Tetracycline
Theophylline	Theo-dur	Elixophyllin, Slo-Phyllin
Triamterene	Dyrenium	None
Trimethoprim/ Sulfamethoxazole	Bactrim	Bactrim DS
Valproic acid	Depakene	None
Warfarin sodium crystalline	Coumadin sodium	Coufarin, Panwarfin

TABLE 11.16 HEALTH PROBLEMS FOR WHICH PARTICULAR DRUGS ARE PRESCRIBED

Generic Name	Indications for the drug
Aluminum hydroxide gel	Antacid
Amiloride HCL	Potassium-sparing diuretic
Aspirin	Arthritis; recurrent TIA, headache

**TABLE 11.16 HEALTH PROBLEMS FOR WHICH PARTICULAR DRUGS
ARE PRESCRIBED (CONT.)**

Generic Name	Indications for the drug
Atenolol	Management of high blood pressure
Carbidopa-Levodopa	Management of Parkinson's disease
Chlorpromazine HCL	Tranquilizer
Chlorpropamide	Management of diabetes
Cholestyramine	Management of high blood cholesterol
Cimetidine	Peptic ulcer
Clonidine HCL	Management of high blood pressure
Colchicine	Antigout; laxative
Diazepam	Tranquilizer
Digoxin	Management of heart disease
Furosemide	Diuretic
Griseofulvin	Antifungal agent
Guanethidine sulfate	Management of high blood pressure
Hydralazine	Management of high blood pressure
Hydrochlorothiazide	Management of high blood pressure
Isoniazid (INH)	Treatment of tuberculosis
Isotretinoin	Treatment for acne
Lithium carbonate	Management of depression and other mental illness
Methotrexate	Anticancer drug
Methyldopa	Management of high blood pressure
Mineral oil	Laxative
Moxalactam disodium	Antibiotic
Neomycin sulfate	Antibiotic
Nitrofurantoin	Treatment of urinary infection
Nitroglycerine	Antiangina medication (for heart disease)
Norethindrone	Oral contraceptive
Norethindrone Mestranol	Oral contraceptive
Penicillamine	Management of arthritis
Phenelzine	Antidepressant
Phenolphthalein	Laxative
Phenytoin	Used for control of seizures
Potassium chloride	Potassium replacement
Prazosin HCL	Management of high blood pressure
Prednisone	Used in severe allergic states
Propranolol	Management of high blood pressure
Ranitidine	Peptic ulcer treatment
Reserpine	Management of high blood pressure
Spironolactone	Management of high blood pressure
Sulfasalazine	Used in ulcerative colitis treatment
Tetracycline	Antibiotic
Theophylline	Used in asthma treatment
Triamterene	Diuretic

TABLE 11.16 HEALTH PROBLEMS FOR WHICH PARTICULAR DRUGS ARE PRESCRIBED (CONT.)

Generic Name	Indications for the drug
Trimethoprim	For urinary tract infections
Valproic acid	For treatment of seizures
Warfarin sodium	Anticoagulant

TABLE 11.17 KEY REFERENCES

Generic	References
Aluminum hydroxide gel	Lotz et al. 1968
Amiloride HCL	Facts and Comparisons 1987
Aspirin	Facts and Comparisons 1987; Lawrence et al. 1984
Aspirin-containing antacid	Facts and Comparisons 1987
Atenolol	Viswanathan and Welling 1984
Carbidopa-levodopa	Facts and Comparisons 1987; Granerus et al. 1977
Chlorpromazine HCL	Lasswell et al. 1984
Chlorpropamide	Tallarida 1984
Cholestyramine	Hashim and Van Itallie 1963; West and Lloyd 1975; Roe 1982
Cimetidine	Tallarida 1984; McCarthy 1983
Clonidine HCL	Tallarida 1984; Facts and Comparisons 1987
Colchicine	Race et al. 1970
Diazepam	Tallarida 1984; Facts and Comparisons 1987
Digoxin	Roe 1982
Furosemide	Roe 1984
Griseofulvin	Facts and Comparisons 1987; Crounse 1961
Guanethidine sulfate	Facts and Comparisons 1987
Hydralazine	Tallarida 1984; Ludden et al. 1982; Facts and Comparisons 1987
Hydrochlorothiazide	Andersson 1984; Grimm et al. 1981; Tallarida 1984
Isoniazid (INH)	Bengoa et al. 1983; Facts and Comparisons 1987
Isotretinoin	Facts and Comparisons 1987
Lithium carbonate or citrate	Facts and Comparisons 1987
Methotrexate	Roenigk and Maibach 1983
Methyldopa	Facts and Comparisons 1987; Myhre et al. 1982

TABLE 11.17 KEY REFERENCES (CONT.)

Generic	References
Mineral oil	Curtis and Ballmer 1939; Fingle and Freston 1979
Moxalactam disodium	Pakter et al. 1982; Facts and Comparisons 1987
Neomycin sulfate	Facts and Comparisons 1987; Faloon 1966
Nitrofurantoin	Rosenberg and Bates 1976
Nitroglycerine	Facts and Comparisons 1987
Norethindrone Ethinylestradiol	See Ortho-Novum 1/50–21
Norethindrone	Tallarida 1984; Roe 1976
Penicillamine	Bergstrom et al. 1981
Phenelzine	Facts and Comparisons 1987
Phenolphthalein	Roe 1976; Heizer et al. 1968
Phenytoin	Tallarida 1984; Reynolds et al. 1966
Potassium chloride liquid	Facts and Comparisons 1987; Palva 1972
Prazosin HCL	Facts and Comparisons 1987
Prednisone	Facts and Comparisons 1987
Propranolol	Melander et al. 1977; Tallarida 1984
Ranitidine	McCarthy 1983; Somogyi and Gugler 1982
Reserpine	Facts and Comparisons 1987
Spironolactone	Facts and Comparisons 1987
Sulfasalazine	Facts and Comparisons 1987; Halsted et al. 1981; Swinson et al. 1981
Tetracyline	Tallarida 1984; Neuvonen 1976
Theophylline	Facts and Comparisons 1987; Feldman et al. 1980
Triamterene	Facts and Comparisons 1987; Roe 1984
Trimethoprim/Sulfamethoxazole	Tallarida 1984; Roe 1981
Valproic acid	Ohtani et al. 1982
Warfarin sodium crystalline	Bjornsson 1984

REFERENCES

ANDERSON, O. 1984. The use of diuretics in modern anithypertensive therapy. *Acta Pharmacol. Toxicol.* Suppl. 54. 1: 79–83.

BENGOA, J.M., M.J.G. BOLT, and I.H. ROSENBERG. 1983. Hepatic vitamin D^{25}-hydroxylase inhibition by cimetidine and isoniazid. Abstract. *Gastroenterology* 84: 1363.

BERGSTROM, R.F., D.R. KAY, T.M. HARKCOM, and J.G. WAGNER. 1981. Penicillamine kinetics in normal subjects. *Clin. Pharmacol. Therap.* 30: 404–31.

BJORNSSON, T.D. 1984. Vitamin K and vitamin K antagonists. In *Drugs and Nutrients: The Interactive Effects,* ed. D.A. Roe and T.C. Campbell. 429–73. New York: Marcel Dekker.

CROUNSE, R.G. 1961. Human pharmacology of griseofulvin: The effect of fat intake on gastrointestinal absorption. *J. Invest. Dermatol.* 37: 529–32.

CURTIS, A.C. and R.S. BALLMER. 1939. The prevention of carotene absorption by liquid petrolatum. *J. Am. Med. Assoc.* 113: 1785–88.

FACTS AND COMPARISONS DIVISION. 1987. St. Louis: J.B. Lippincott.

FALOON, W.W. 1966. Effect of neomycin and kanamycin upon intestinal absorption. *Ann. N.Y. Acad. Sci.* 132: 879–87.

FELDMAN, C.H., V.E. HUTCHINSEN, C.E. PIPPENGER, J.A. BLUMENFELD, B.R. FELDMAN, and W.J. DAVIS. 1980. Effect of dietary protein and carbohydrate on theophylline metabolism in children. *Pediatrics* 66: 956–62.

FINGL, E. and J.W. FRESTON. 1979. Antidiarrhoeal agents and laxatives: changing concepts. *Clin. Gastroenterol.* 8: 161–86.

GRANERUS, A.-K., R. JAGAENBURG, and A. SVANBORG. 1977. Kaliuretic effects of L-dopa treatment in Parkinsonian patients. *Acta Med. Scand.* 201: 291–97.

GRIMM, R.H., A.S. LEON, D.B. HUNNINGHAKE, K. LENZ, P. HANNAN, and H. BLACKBURN. 1981. Effects of thiazide diuretics on plasma lipids and lipoproteins in mildly hypertensive patients. *Ann. Intern. Med.* 94: 7–11.

HALSTED, C.H., G. GANDHI, and T. TAMURA. 1981. Sulfasalazine inhibits the absorption of folates in ulcerative colitis. *New Engl. J. Med.* 305: 1513–16.

HASHIM, S.A. and T.B. VAN ITALLIE. 1963. Experimental steatorrhea in human subjects. In *Malabsorption Syndromes.* 26–30. Basel: Karger.

HEIZER, W.D., A.L. WARSHAW, T.A. WALDMAN, and L. LASTER. 1968. Protein losing gastroenteropathy and malabsorption associated with factitious diarrhea. *Ann. Intern. Med.* 68: 839–52.

LASSWELL, A.B., D.A. ROE, and L. HOCHHEISER. 1985. *Nutrition for Family and Primary Care Practitioners.* Philadelphia: George F. Stickley.

LAWRENCE, V.A., J.E. LOEWENSTEIN, and E.R. EICHNER. 1984. Aspirin and folate binding: in vivo and in vitro studies of serum binding and urinary excretion of endogenous folate. *J. Lab. Clin. Med.* 103: 944.

LOTZ, M., R. NEY, and F.C. BARTTER. 1964. Osteomalacia and debility resulting from phosphorus depletion. *Trans. Assoc. Am. Physicians* 11: 281.

LOTZ, M., E. ZISMAN, and F.C. BARTTER. 1968. Evidence for a phosphorus depletion syndrome in man. *New Eng. J. Med.* 273: 409–15.

LUDDEN, T.M., J.L. McNAY, Jr., A.M.M. SHEPHERD, and J.M.S. LIN. 1982. Clinical pharmacokinetics of hydralazine. *Clin. Pharmacokinet.* 7: 185–205.

McCARTHY, D.M. 1983. Editorial. Ranitidine or cimetidine. *Ann. Intern. Med.* 99.

MELANDER, A., K. DANIELSON, B. SCHERSTEN, and E. WAHLIN. 1977. Enhancement of the bioavailability of propranolol and metoprolol by food. *Clin. Pharmacol. Ther.* 22: 108–12.

MYHRE, E., H.E. RUGSTAD, and T. HANSEN. 1982. Clinical pharmacokinetics of methyldopa. *Clin. Pharmacokinet.* 7: 221–33.

NEUVONEN, P.J. 1976. Interactions with the absorption of tetracycline. *Drugs* 11: 46–54.

OHTANI, Y. 1982. Carnitine deficiency and hyperammonemia associated with valproic acid therapy. *J. Pediat.* 101: 782–85.

PAKTER, R.L., T.R. RUSSELL, H. MIELKE, and D. WEST. 1982. Coagulopathy associated with the use of Moxalactam. *J. Am. Med. Assoc.* 248: 1100.

PALVA, I.P., S.J. SALOKANNEL, T. TIMONEN, and H.L. PALVA. 1972. Drug-induced malabsorption of vitamin B_{12}. IV. Malabsorption and deficiency of vitamin B_{12} during treatment with slow release potassium chloride. *Acta Med. Scand.* 191: 355–57.

POWERS, D.E. and A.O. MOORE. 1986. *Food-Medication Interactions.* Phoenix, AZ: FMI.

RACE, T.F., I.C. PAES, and W.W. FALOON. 1970. Intestinal malabsorption induced by oral colchicine. Comparison with neomycin and cathartic agents. *Am. J. Med. Sci.* 259: 32–41.

REYNOLDS, E.H., G. MILNER, D.M. MATTHEWS, and I. CHANARIN. 1966. Anticonvulsant therapy, megaloblastic haemopoiesis, and folic acid metabolism. *Quart J. Med.* [N.S.] 35: 521–37.

ROE, D.A. 1976. *Drug-Induced Nutritional Deficiencies.* Westport, CT: AVI.

———. 1981. Drug interference with the assessment of nutritional status. *Clin. Lab. Med.* 1: 647–64.

———. 1982. Drug-nutrient interrelationships. *Pract. Gastroenterol.* 6: 32–38.

———. 1984. Drug-induced mineral depletion in the elderly. In *Drugs and Nutrition in the Geriatric Patient.* ed. D.A. Roe. 105–20. New York: Churchill Livingstone.

———. 1984. Therapeutic significance of drug-nutrient interactions in the elderly. *Pharmacol. Rev.* 36: 109S–122S.

———. 1985. Therapeutic effects of drug-nutrient interactions in the elderly. *J. Am. Diet. Assoc.* 85: 174–81.

ROENIGK, H.H. and H.I. MAIBACH. 1983. Methotrexate. *Seminars in Dermatology* 2: 231–37.

ROSENBERG, H.A. and T.R. BATES. 1976. The influence of food on nitrofurantoin bioavailability. *Clin. Pharmacol. Ther.* 20: 227–30.

SOMOGYI, A. and R. GUGLER. 1982. Drug interactions with cimetidine. *Clin. Pharmacokinet.* 7: 23–41.

SWINSON, C.M., J. PERRY, M. LUMB, and A.J. LEVI. 1981. Role of sulfasalazine in the aetiology of folate deficiency in ulcerative colitis. *Gut* 22: 456–61.

TALLARIDA, R.J. 1984. *Top 200. The most widely prescribed drugs in America.* 17–18, 46–47, 66, 190–91, 196–97, 213–14. Philadelphia: W.B. Saunders.

U.S. PHARMACOPOEIAL CONVENTION, INC. 1986. *Advice for the Patient: Drug Information in Lay Language,* vols. 1 and 2. Rockville, MD: U.S. Pharmacopoeial Convention.

VISWANATHAN, C.T. and P.G. WELLING. 1984. Food effects on drug absorption in the elderly. In *Drugs and Nutrition in the Geriatric Patient.* ed. D.A. Roe. 47–70. New York: Churchill Livingstone.

WEST, R.J. and J.K. LLOYD. 1975. The effect of cholestyramine on intestinal absorption. *Gut* 16: 93.

INDEX

215